Beyond the Limits

The MIT Press
Cambridge, Massachusetts
London, England

Beyond the Limits

Flight Enters the Computer Age

Paul E. Ceruzzi

This book was set in Helvetica Condensed and
Bookman by Achorn Graphic Services, Inc. and
printed and bound by Arcata Graphics/Kingsport
in the United States of America.

Library of Congress Cataloging-in-Publication
Data
Ceruzzi, Paul E.
  Beyond the limits.
  Bibliography: p.
  Includes index.
    1. Aeronautics—Data
processing.   2. Astronautics—Data
processing.   I. Title.
TL553.C38   1989   629.13′0028   88-23191
ISBN 0-262-03143-4
ISBN 0-262-53082-1 (pbk.)

To Carla

Contents

Preface

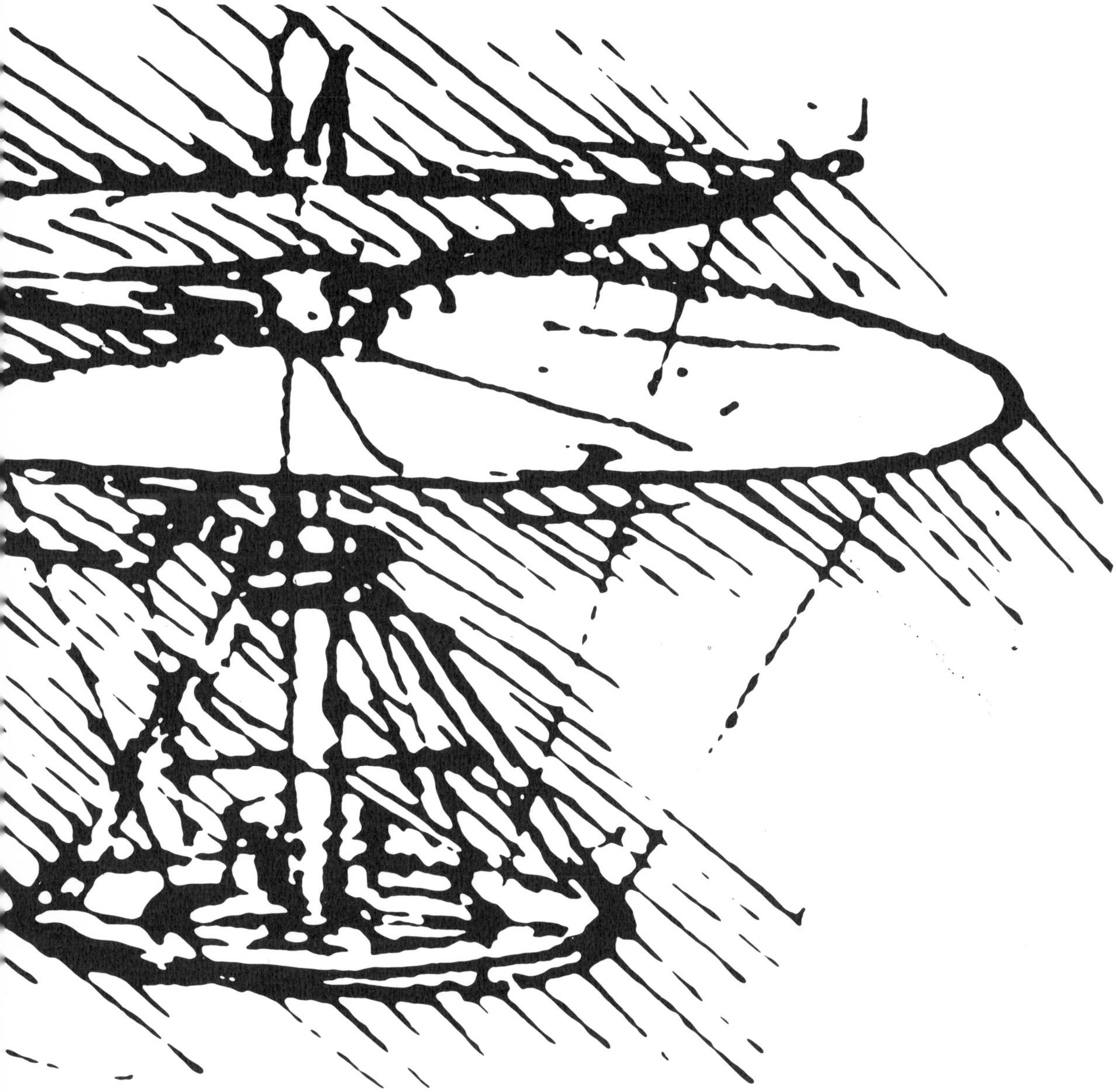

In the fifteenth century Leonardo Da Vinci sketched a number of possible configurations for flying machines in his notebooks. The advent of powered flight would have to wait for a more thorough knowledge of aerodynamics and more advanced power plants than were available in Da Vinci's day. (Smithsonian Institution)

The age-old dream of flight in this century has been achieved through the skillful application of mind, but not mind alone. From the beginning of human attempts at flight, men and women turned to mechanical devices to augment the powers of the human brain as they grappled with problems of lift, propulsion, and control of flying machines. These instruments were vital to the design, testing, navigation, and control of airplanes and rockets (see figure).

During and immediately after the Second World War, the demands of designing and building supersonic, jet-propelled aircraft and ballistic missiles overwhelmed the computing tools of an earlier era. To meet those demands, aircraft and rocket engineers turned to a new tool, the electronic digital computer. The computer was itself invented in the 1940s in response to a number of wartime projects, some having little to do with flight technology. But as news of its invention spread, the community of air and space engineers, scientists, and pilots were among the first to recognize its value for their own work. It was not easy to get the early computers to work, but those men and women became enthusiastic supporters of its development.

Since then, advances in computer technology have kept pace with advances in aerospace. Nowadays computers of all types, sizes, and shapes help design, test, manufacture, navigate, and operate flying machines. Computer simulations train pilots and astronauts, and computers play a key role in understanding the science of aerodynamics.

This book was written to coincide with the opening of a gallery of the same name at the Smithsonian Institution's National Air and Space Museum in Washington, D.C. The gallery had its origins in a desire to update an existing exhibit on flight technology in a gallery that opened in the museum in July 1976. Early in the planning for this new exhibit its curators recognized that the computer more than anything else had changed flight technology since 1976. The dramatic shrinking of computing power onto microscopic chips of silicon dovetailed with aerospace's needs not only for that power but also for machines that are lightweight, compact, and rugged enough to withstand the temperature and vibration extremes of the flight environment. The result was a thorough penetration of computing into every aspect of flight technology, to a point where today one cannot speak of modern flight without talking of its dependence on computers.

So-called revolutions in technology, when they occur, often seem to happen suddenly with little warning. But this change in aerospace has been going on for at least forty years, since the end of the Second World War. Almost as soon as the first electronic computers began operating in the late 1940s, aircraft engineers sought to make good use of them. Those early efforts laid the foundation for the more rapid changes of the past decade. How this change came about and what its consequences are, are the subject of this book.

Acknowledgments

This book would have been impossible without the gallery at the National Air and Space Museum, from which it sprang. An author of a book may acknowledge the contributions of others; still the preparation of a book is by and large a solitary job. That is not so with a gallery, which is much more akin to opening a Broadway play than to writing a book. A Smithsonian exhibit involves dozens of people. They are as much a part of this narrative as they are a part of the exhibition. William "Jake" Jacobs, the gallery's designer, and Jim O'Neill, the gallery's computer specialist, deserve special recognition for their tireless and inspired efforts over the long time this project has been in the making. Harold Motin, Edwin Collen, and Jack Van Ness helped all along. I wish also to thank those others at the museum, especially those in the front office, who supported this project through its many ups and downs. I am especially grateful to Marilyn Lyons, Susan Beaudette, and John Carlin, who worked so hard getting support for the gallery at times when it was little more than a few words on a piece of paper.

A generous grant from Cray Research helped underwrite the cost of producing this book. I wish also to thank those organizations that have joined the Smithsonian in sponsoring the gallery: Digital Equipment Corporation, Intelsat, Prime Computer, Hitachi America, Bell Atlantic Corporation, Molex, the Unisys Corporation, General Dynamics Corporation, Next, Northrop Corporation, and Hughes Aircraft Company. Apple Computers organized and underwrote the Race for Space software contest, the winning entries of which are featured in the gallery.

The basic subject matter of the gallery and this book was hammered out during long discussions among people of the computer and aerospace industries, historians, and museum professionals. I especially wish to thank William Howard of Coopers and Lybrand, Jay Barcheski of Gould, Les Rosenthal of Digital Equipment Corporation, and Tom Kastner of Grumman Corporation for their initial advice and encouragement . for this project. At the museum Allan Needell, Lou Lomax, Nan Knight, Vic Govier, and Howard Wolko performed a similar service.

Several readers of the manuscript pointed out errors of fact, sloppy prose, and logical inconsistencies, and to them I am also grateful: Martin Campbell-Kelly, William Aspray, James Hansen, James Tomayko, John Norton, and James Sharp. Mike Williams, Jon Eklund, Marie Mattson, and David Allison, all of the Smithsonian's Museum of American History, lent support and gave valuable advice throughout the project.

Beyond the Limits

Chapter 1

The Evolution of Computers and Aerospace

How does one explain the history of something as complicated as air and space technology? And how does one fit into that history the role of the computer, itself a machine of enormous subtlety and complexity? Historians look for general trends, patterns, and statements that at least try to explain *why* things look the way they do. But the complexity of modern technology makes that hard to do. Too often the histories of aerospace technology lack any organizing structure and thus become little more than long lists of successive machines with accompanying lists of their increasingly better features. In looking at the story of computing and flight, I have adopted a structure commonly used in history, but I have changed it slightly to accommodate the unusual nature of this particular story. That structure is not the only way to organize this material. But it works, so long as one uses it carefully and keeps in mind that it is an aid to understanding and not a "law" of history.

These complex machines have a counterpart in nature; namely, complex living things. For that reason observers of modern technology have often borrowed terms from evolutionary biology to explain its dynamics. Writers often classify airplanes, spacecraft, and computers as belonging to successive "generations." They describe modern machines as having "evolved" from earlier variants, implying a Darwinian selection of features and designs from earlier models and a dying out of other, less optimal designs that somehow did not fit the economic or military environment in which they had to survive.

This model does not strictly apply to artifacts designed and made by humans. Like living things, machines are complex, but unlike the natural environment, engineers consciously design machines to serve specific ends. They are free to borrow specific features from many different parents for a technical innovation. And unlike natural species, machines can have more than two parents. The Wright brothers' 1903 Flyer had as its parents the technologies of railroad bridges, from which the structure of the wing box came; automobiles, from which the engine came; and the bicycle, from which the ideas of balance and control came. Human ingenuity created Pegasus, the beast that combined the wings of a bird with the power of a horse, but in the natural world no such animal with multiple parenthood could exist. Nevertheless, if applied with caution, the biological metaphor can help us understand the history of a technology.

Understanding the relationship of computing technology to the history of air and space craft benefits from carrying this analogy a step further. Since 1945 the evolution of these two technologies has been greatly intertwined. At various times advances in each have led directly to advances in the other. At other times the technical, social, or financial needs of one were met by the other. The state of computing technology meant that aerospace could not take the computer industry's products as they were; aerospace's own special needs as well as financial, political, and technical resources made it an active participant.

Nature knows many examples of a mutual, rapid coevolution of two disparate forms of life, each dependent on the other for its existence and its characteristics. Flowering plants, for example, appeared on the earth at the same time as winged insects; it is hard to conceive of the emergence of colorful flowers without a parallel emergence of forms of life that benefited from them. In other cases, animals or plants suitable as food for a predator developed mechanisms for defense (such as camouflage); these mechanisms often had to evolve rapidly as predators responded with a rapid evolution of their hunting skills.

Such an interaction characterizes the development of computing and aerospace technology after the Second World War. The technology of flight has lived in a symbiosis with that of computing. Yet it is more than a symbiosis, which is usually a static, steady state of affairs. The interaction of the two technologies has propelled each to evolve much faster than each could possibly have evolved separately. Many facets of modern air and space flight would be impossible without computing; the current state of computing would be far less advanced without aerospace.

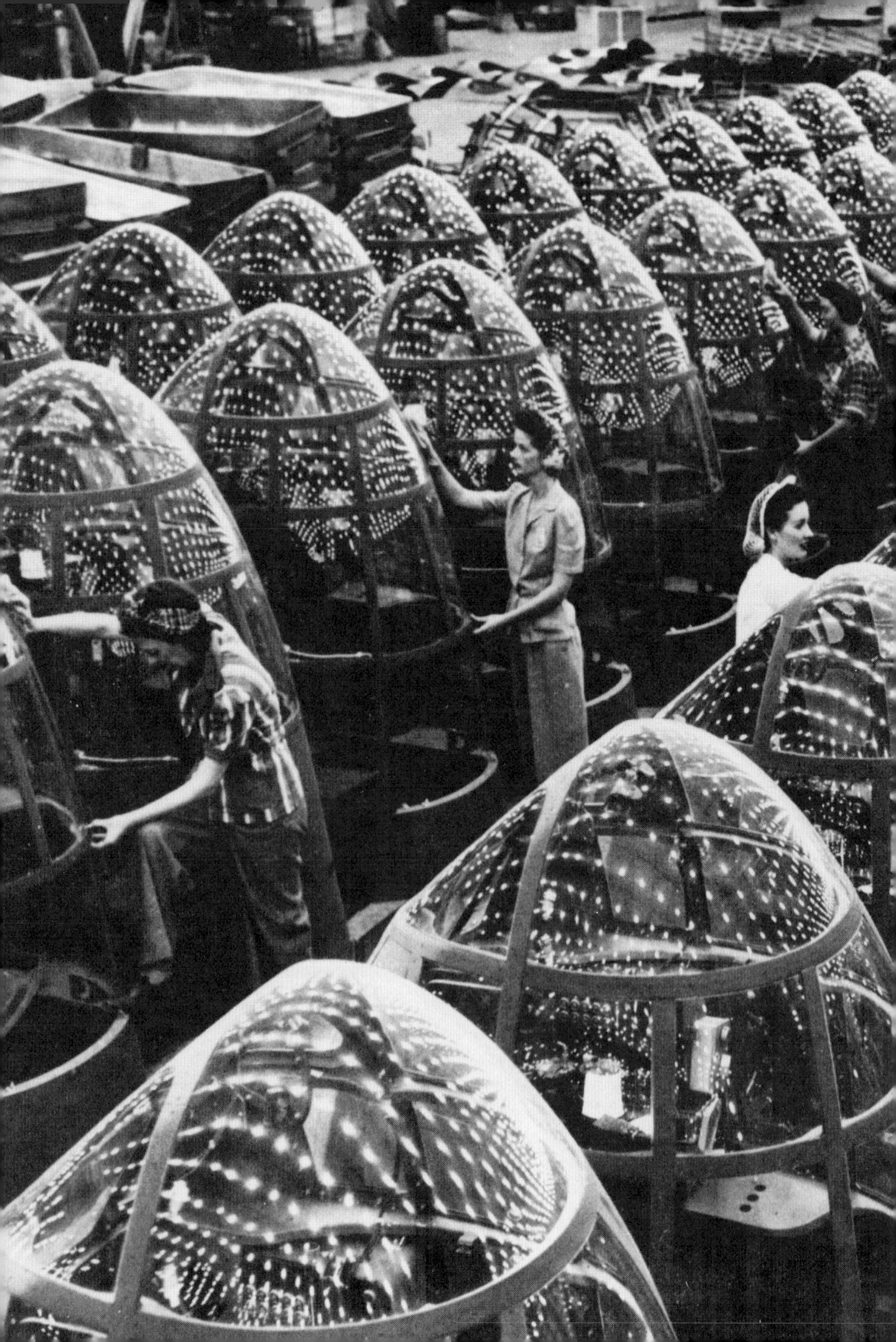

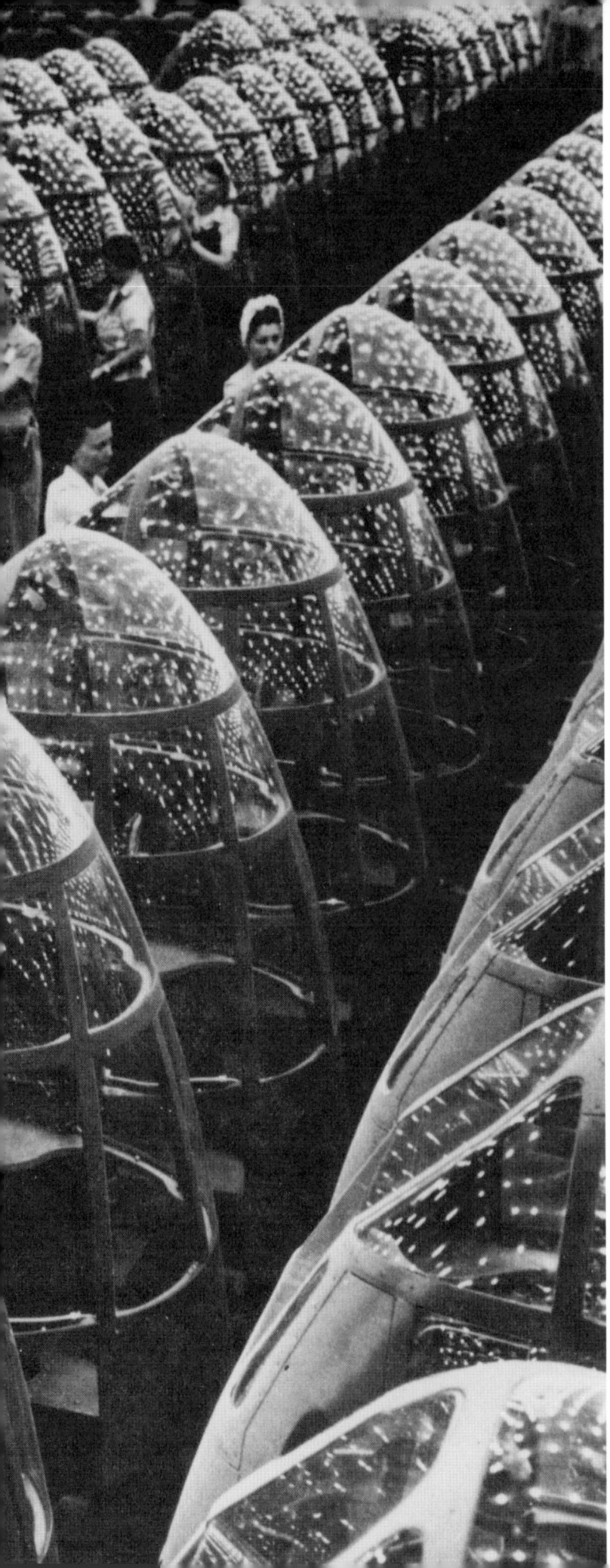

1.1
Workers at the Douglas Aircraft Company's plant in Long Beach, California, are assembling the plexiglas canopies for an American warplane around 1943. The canopies reflect the pattern of lights overhead. (Photo: Library of Congress)

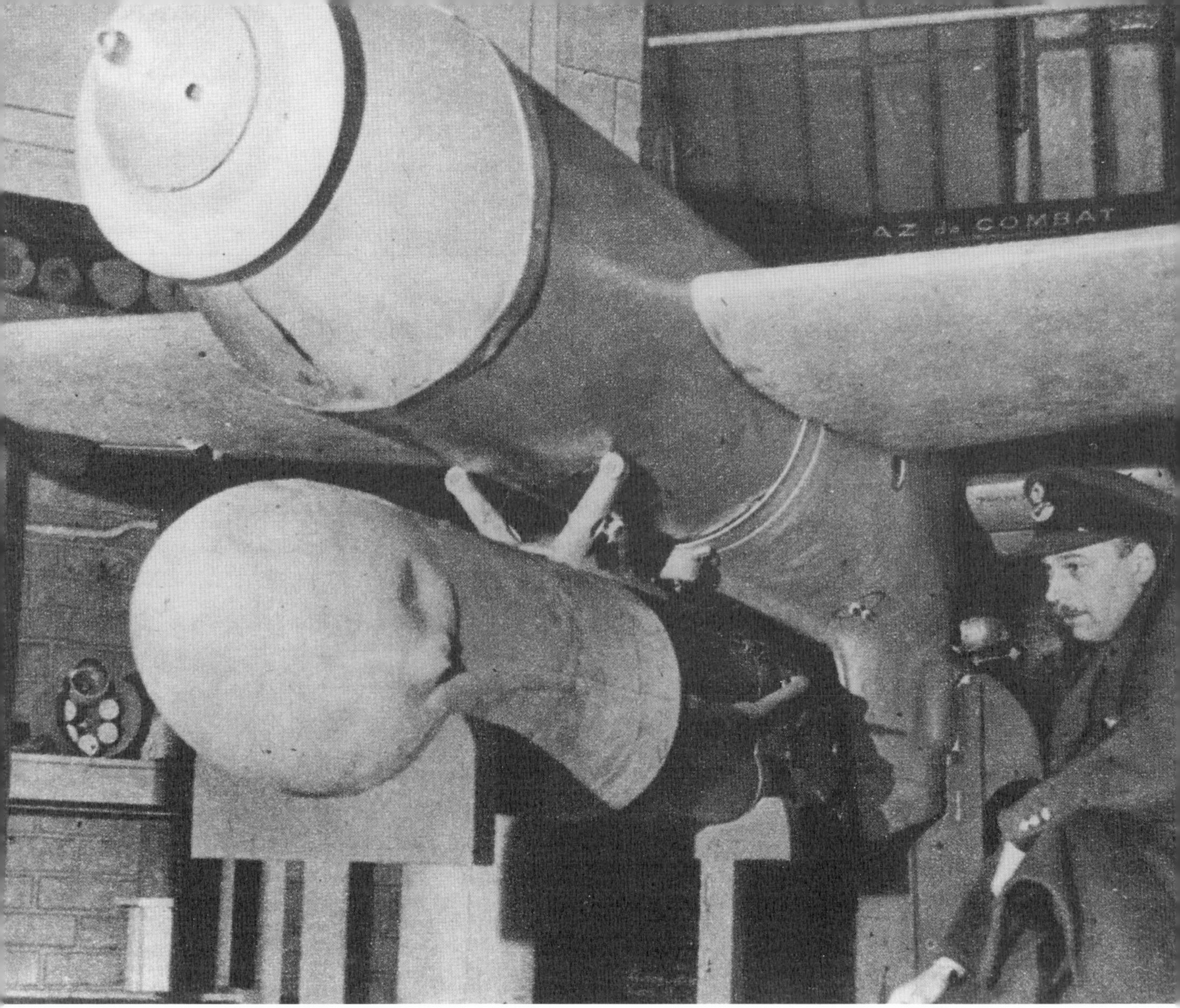
AZ de COMBAT

a

1.2

HS-293. Late in World War II the German Air Force deployed this radio-guided flying bomb (*a*). Dropped from an airplane, it was guided to its target by the bomber crew, who could thus remain above the level of enemy antiaircraft fire. The mass production of the HS-293 wing (*b*) was assisted by a computer invented by Konrad Zuse. (Photos: National Air and Space Museum and David E. Dempster)

Aerospace from 1945 to 1955

Let us first consider aeronautic technology as it existed in the United States in 1945. Few would deny the importance of aircraft to the Allied victory over Germany and Japan in the Second World War. For the United States, the victory was one of production; America's hero was Rosie the Riveter, who helped turn out airplanes in unprecedented numbers (figure 1.1). In 1939 President Roosevelt asked for 30,000 warplanes. By 1944 American factories were turning out that many airplanes every four months. Production was the Allies' strength; it was enough to overcome serious weaknesses in America's technological base, which otherwise might have spelled disaster.

As the Allied armies overran Germany, reports came back confirming earlier suspicions that the United States had fallen behind in crucial areas of aircraft technology. The Germans had made little progress in the development of an atomic bomb, but they were far ahead in the development of jet and rocket propulsion, swept-wing aircraft, and other design innovations that pointed toward supersonic flight. Germany had developed antiaircraft rockets and advanced weapons like the Henschel HS-293, an unmanned flying bomb released from an airplane and radio-guided to its target (figure 1.2). The nature of aeronautics was changing.

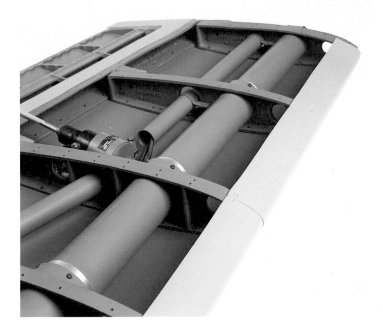

b

In rocketry, the American Robert Goddard had been making
steady progress in liquid-fueled rockets, but his work paled
beside the achievements of Wernher von Braun and his
Peenemünde rocket team in rocket technology (figure 1.3).
In 1945 von Braun and some of his team were brought to the
United States, where the U.S. Army established a modest program using captured V-2 rockets. At the White Sands test
range in New Mexico, the V-2 firings achieved higher and
higher altitudes, breaching the earth's atmosphere and entering a new realm, which von Braun's group called outer space.

The quick emergence of the cold war with the Soviets and
their acquisition of nuclear weapons only a few years after
Hiroshima intensified the sense of urgency and change within
the American aircraft industry, the Air Force (separated from
the Army in 1947), the Army, and the Navy. The Soviets began to vigorously develop rocket technology at the end of the
war, and by 1951 they were ahead of the United States in the
design of rockets that could lift heavy payloads into space. The
Soviet triumph of Sputnik in October 1957 came as a shock to
most Americans, yet it was a triumph based on a consistent
program of government support for rocket technology in place
for at least a decade.

In the United States the shock of Sputnik led to rethinking
the relationship between space exploration and aeronautics.
Airplanes are in many ways different from missiles. Technical
issues lent support to either keeping the two activities separate or consolidating them. Rivalries among the U.S. armed
services and the intelligence community added a political dimension to the question. But after the Soviet triumph of
Sputnik, the view of the Air Force prevailed: for military purposes there are no fundamental differences between air and
space. Air Force spokesmen coined the term "aerospace" to
reflect and help establish the perception that what air and
space have in common is more important than what make
them different.[1]

1.3

**V-2 rockets in Europe, ca. 1945. (Photo: National
Air and Space Museum)**

Computing from 1945 to 1955

The postwar era saw a transformation of computing technology that matched the transformation of aeronautics. The

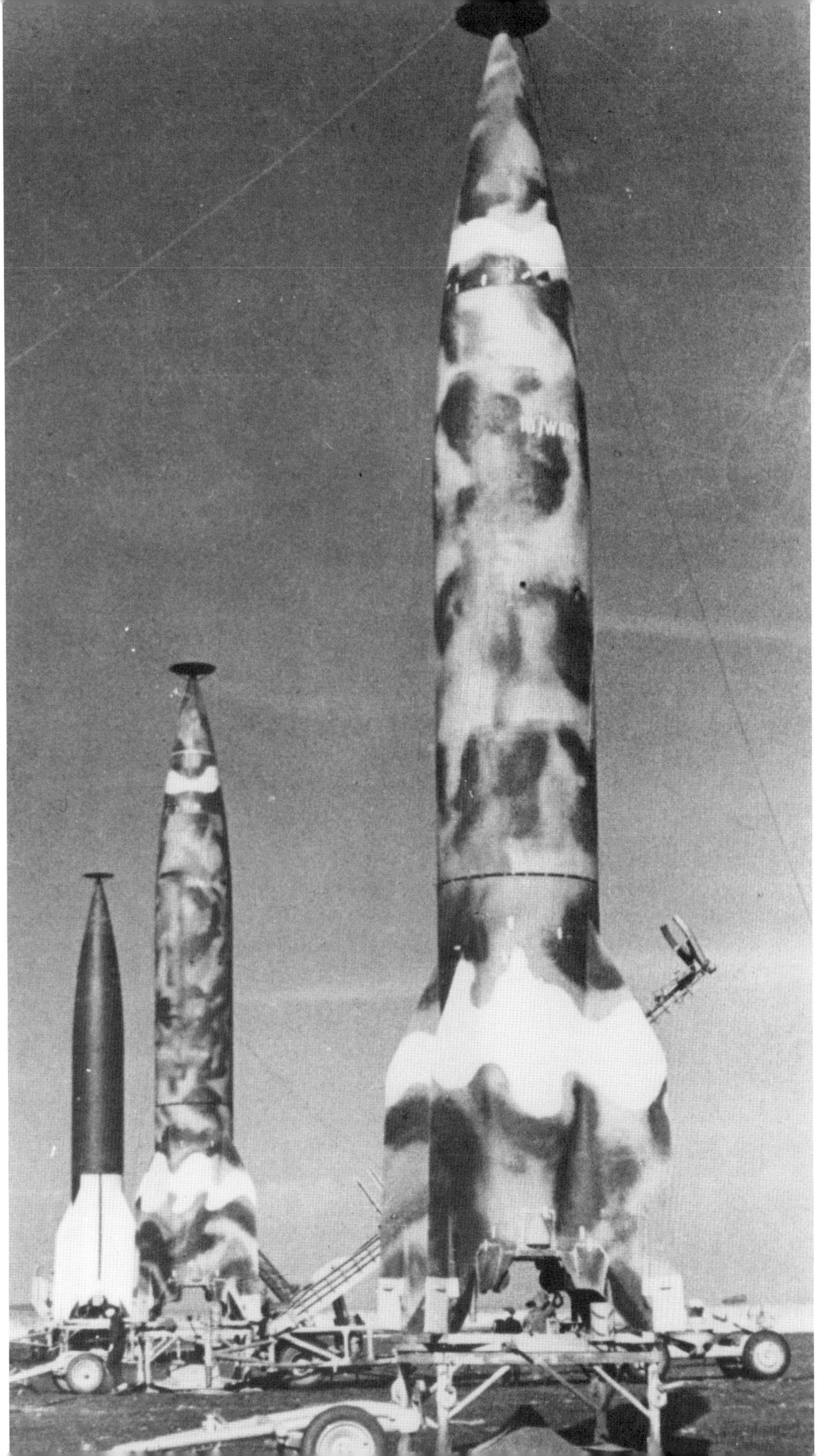

digital computer, able to perform automatically a variety of arithmetic, memory, and control functions, was invented in the 1940s. After an initial period of slow growth, a commercial market developed and eventually experienced an explosive growth unmatched in the history of technology. In 1945 only about four or five machines in the world could automatically perform complex sequences of calculations for scientific or engineering work. By 1955 hundreds existed; ten years later, thousands. By the 1980s computing machines were so numerous that any attempt to number them is hopeless.

The idea for such a machine, as well as the technical means for realizing it, existed a hundred years before the first computers were built. The English mathematician Charles Babbage sketched out a plan for a fully automatic digital computer in the 1830s. Although the technology of calculating machines advanced steadily from Babbage's day, his idea of a fully automatic computer lay dormant for many years prior to the breakthroughs of the late 1940s. Many historical accounts argue that the technical level of machine tooling and metallurgy prevented Babbage from realizing his design. Babbage's 1838 design for an "analytical engine" called for closer tolerances, better finish, and more uniform metal properties than any machines of his day. But Babbage's designs were feasible at the state of British machine techniques of his day.[2] And the technology of electromagnetic relays and vacuum tubes, from which the first computers of the 1940s were built, was available and widespread certainly by the 1920s.

The advent of war in 1939 changed the picture. The immediate effect of the war was to provide support and funds for, and a sense of urgency to, a number of ideas for automatic computers, including a few that had reached a "breadboard" or prototype stage by 1939. All computer projects since Babbage's day lacked these elements; they made his dream come true.

In Germany the German Aeronautics Research Bureau supported a modest effort by an aircraft engineer named Konrad Zuse to help with an aircraft design problem known as flutter (this problem will be addressed in a later chapter). In England the British Foreign Office designed an electronic computer to unscramble intercepted German radio messages. In the United States the military heavily supported at least three separate projects: one at Harvard University, another at Bell Telephone Laboratories, and a third at the University of Pennsylvania's Moore School of Electrical Engineering. Those computers were used for a number of purposes but mainly to calculate trajectories of heavy artillery and antiaircraft guns.

The Moore School project had the most influence on the direction of computing technology. The machine produced there, ENIAC, calculated with electronic tubes, which made the machine many times faster than the mechanical devices produced elsewhere. Completed just after the end of the war in 1945, ENIAC (an acronym for "Electronic Numerical Integrator and Computer") became the first operational electronic digital computer and marked the beginning of generations of machines continuing to the present day.

Of all the wartime computing projects only that of Zuse in Germany was directly supported by an aeronautics organization, and only it was explicitly designed for an aeronautics application. None of the American computing projects were driven by a need in aeronautics or missile technology. (Contrary to folklore, digital computers contributed almost nothing to the famous secret weapons of the war, including the atomic bomb, the V-2, and radar.) But as news of the new invention spread after the war's end, the community of aeronautic engineers and scientists recognized its potential. It offered precisely the help they needed in managing the complexity of the postwar technology.

The aerospace community became a key agent in bringing the digital computer out of the laboratory and into general commercial use and now personal use. The computing needs for ballistics, code-breaking, and atomic weapons development remained strong after 1945, but they did not match the needs or financial resources of those involved with aeronautics and missiles. This group included engineers designing jet aircraft, test pilots and ground crews proving out these new designs, and finally the Army, Navy, and Air Force, each trying to fit aircraft and guided and ballistic missiles into their organizations.

The first electronic computers were expensive and troublesome to operate and maintain. Some of them could not operate for more than a few minutes at a time before a malfunction would occur. Many who saw the potential of the computer decided to wait until things settled down a bit before abandoning less automated methods that at least worked. Potential customers had to be familiar with the engineering of large electrical and mechanical systems, and they had to put up with the mechanical breakdowns and difficulties involved in getting those machines to yield useful results. Government agencies like the U.S. Census Bureau and commercial customers like insurance companies had a need for automatic computing power and were among the first to purchase or lease versions of the new invention. But they lacked the in-house engineering savvy to operate and maintain these machines, so they often failed to make good use of the early computers. In some instances these customers turned to government agencies like the Bureau of Standards to do their engineering for them. While this approach worked, it put the customer at the mercy of an outside agency. Because of political or financial difficulties, that agency often could not deliver what the customer needed.

Aerospace was different—it was accustomed to complex machines that needed long break-in periods. Unlike those in the insurance companies or banks, its employees were comfortable with electronic systems, a legacy of World War II radar and communications systems that went into warplanes. It had ample financial resources from Congress and the newly created Department of Defense. Those who worked in aerospace understood that the first model of an airplane usually had numerous bugs, which had to be rooted out, and so they were patient with balky equipment.

Finally, aerospace had a critical mass of suppliers, manufacturers, and customers in both the military and commercial sectors. Some aerospace projects were undertaken in secrecy, but unlike nuclear weapons or code-breaking projects, many were not. The aerospace community was large with important concentrations on both coasts; its members included university faculty and students, civil and military government agencies, airframe manufacturers, and commercial airlines. Together they formed a community of users who actively guided and shaped the early computer industry as it grew.

Computing versus Aerospace Technology

One often hears the term "high technology" in the context of contemporary machines that use advanced materials and perform exotic functions hardly envisioned a few decades ago. Modern computers and air and space craft fit this category. But in terms of their nature, the two types of machines could hardly be more different. There is no such thing as a general-purpose flying machine. Air and space craft are designed with a specific application. This is obvious to even an untrained eye. A digital computer, in contrast, is a general-purpose machine whose function is often defined only after it leaves the factory. Depending on the way it is programmed, the same machine may be used for engineering calculations, word processing, calculating a payroll, maintaining a list of names and addresses, or drawing a picture.

This ability comes from its inherent design. First, it computes digitally; that is, in discrete numeric quantities that can be manipulated by the familiar rules of arithmetic. By contrast, an analog computer, which up to 1950 was more common in aeronautics, computes by constructing a physical device whose behavior is similar to the problem to be solved. The familiar analog watch tells time by constructing a miniature model of the earth's daily revolution. The speed of the hands of the watch is set to correspond to the speed of the revolutions of the earth. A digital watch simply counts seconds and displays that count as a set of numbers. Its internal design does not suggest in any way the physical revolutions of the earth, whose time it tells. Analog computers tend to be simpler, and in many cases the problem to be solved suggests an obvious design for the computer. But they have almost no flexibility—one must build a different analog computer for every different type of problem to be solved. The flexibility of a digital computer is one key to its power.

Second, digital computers store their programs of instructions internally in the same memory device they use for storing the numbers involved in solving a problem. That means that the computer can call on those instructions at high speeds and even modify those instructions in response to the calculations it has already performed. A stored program allows a person to take advantage of the inherent flexibility of a digital computer so that it can solve a wide range of problems with few limits. To use the example of the wristwatch again, an analog watch built to tell time on earth would not be suitable for use on another planet where the length of the day is different. All the gears would have to be replaced. A digital watch built for use on earth would do fine, with only a simple change in its program. Moreover, the circuits of a digital watch can also perform computing tasks unrelated to time-keeping like doing arithmetic or storing telephone numbers. All that changes is the internal program.

The external form of computers has stabilized in recent years to a familiar configuration of a rectangular box containing the processor, a modified television screen to display results, and a modified typewriter keyboard to accept commands and data from an operator. (I do not include in this description computers embedded into other machines; these take on the shapes of the host machine and in many cases are all but invisible.) Computers used in aerospace applications, whether on board or on the ground, are in many ways no different from those used in just about any other segment of society. Many effects that the computer has had on aerospace are the same as those it has had on other segments of the economy. The following chapters concentrate on those that are most striking and unique to aerospace: primarily activities connected with the design, aerodynamics, and manufacture of high-performance aircraft and with the navigation and control of manned spaceships.

The next five chapters focus on events in aerospace engineering that transformed flight between 1945 and 1975. Chapters 7 and 8 survey the synergistic effects of aerospace and computer technology after 1975, and chapter 9 looks at computer software for aerospace applications. An appendix discusses the role of computers in the explosion of the Space Shuttle *Challenger* in January 1986.

Chapter 2

Northrop: Midwife of the Computer Industry

At the center of postwar aeronautics and rocket activity was
Northrop Aircraft of Hawthorne, California. This company
was founded in 1939 by John K. Northrop, who by then had
an international reputation as a designer of high-performance
airplanes. In the late 1920s he perfected the technique of cov-
ering the internal metal structure of the wings and fuselage
with a skin made of an aluminum alloy, not doped fabric. This
skin not only covered but also supplied much of the airplane's
structural strength, an innovation that both reduced weight
and made unnecessary the struts and bracing wires that contemporary airplanes then required for structural strength.

The Flying Wing and the Snark

In the early years of the cold war and because of worsening
relations with the Soviet Union, Northrop became involved in
two major projects aimed at carrying a war across the oceans
to Asia or Europe. The first was the Snark guided missile; the
second the Flying Wing manned bomber. Both these craft rep-
resented radical breaks with the traditions of aircraft and
rocket design of the previous decade, and they incorporated a
host of design innovations that would eventually become com-
monplace. Neither craft was a success for Northrop. Neither
is much remembered today. But each pointed to a future in
which flight would be very different from what it was during
World War II. In designing these two craft, Northrop opened
paths that would eventually become highways in flight technology traveled by the Space Shuttle, cruise missiles, and the
Concorde among others. Moreover, computational demands
for the design of these craft made Northrop Aircraft the mid-
wife of the computer industry in the United States.

The Flying Wing was just that, a flying wing (figure 2.1). In its
initial design it had no tail, rudder, or fuselage. (When jet engines were substituted for piston engines as the power source,
small vertical tails were added.) The advantage of that design
was that every piece of the airplane contributed to lift, with
none of the drag added by those other structures. The concept
was sound and offered tremendous performance advantages
over conventional designs. But to implement it required working at the cutting edge of aerodynamic theory at that time. A
combination of flaps and other devices on the wings gave the
pilot the same ability to maneuver and control the craft as on

2.1

After the war Northrop Aircraft Company proposed this design for an Air Force heavy bomber.
The YB-49 Flying Wing had only vestiges of a tail
and fuselage, a design that minimized drag and
theoretically led to a more efficient craft. The
design placed heavy demands on the stability
and control systems, however, and problems
with those systems led to the cancellation of
the project in 1949. (Photo: National Air and
Space Museum)

any conventional airplane, but the unusual configuration required an internal stabilization system that went far beyond the typical autopilots used in conventional bombers. And the Flying Wing was big, which meant its structural engineering had to be done carefully.[1]

The Snark was a subsonic, unmanned, winged missile designed to carry a nuclear warhead from the North American continent to targets in the Soviet Union (figure 2.2). Like the Flying Wing it had no horizontal tail, although it did have a vertical fin and fuselage. Two solid-fueled rockets launched the Snark from a stationary platform; once airborne, the rockets dropped off and a turbojet engine took over. Everything about the Snark—its jet and rocket propulsion, intercontinental mission, swept wings, nuclear warhead, and above all lack of a human pilot—set it apart from the typical bombers of just a few years before.

Without a human pilot the Snark required some sort of intelligence to guide it to a target. At launch an internal radio system measured its velocity and adjusted the craft's controls to keep it on the correct course. But once on its way, radio was not acceptable. It would be out of range while flying over the ocean. And once over enemy territory, it could be jammed. Guidance of the Snark thus had to be self-contained and autonomous for most of its flight. It also had to be lightweight, rugged, and compact. So for all but the initial and final phases of flight, the Snark used a star-tracking system that in many respects was an unmanned version of the way sailors had navigated the oceans for years.

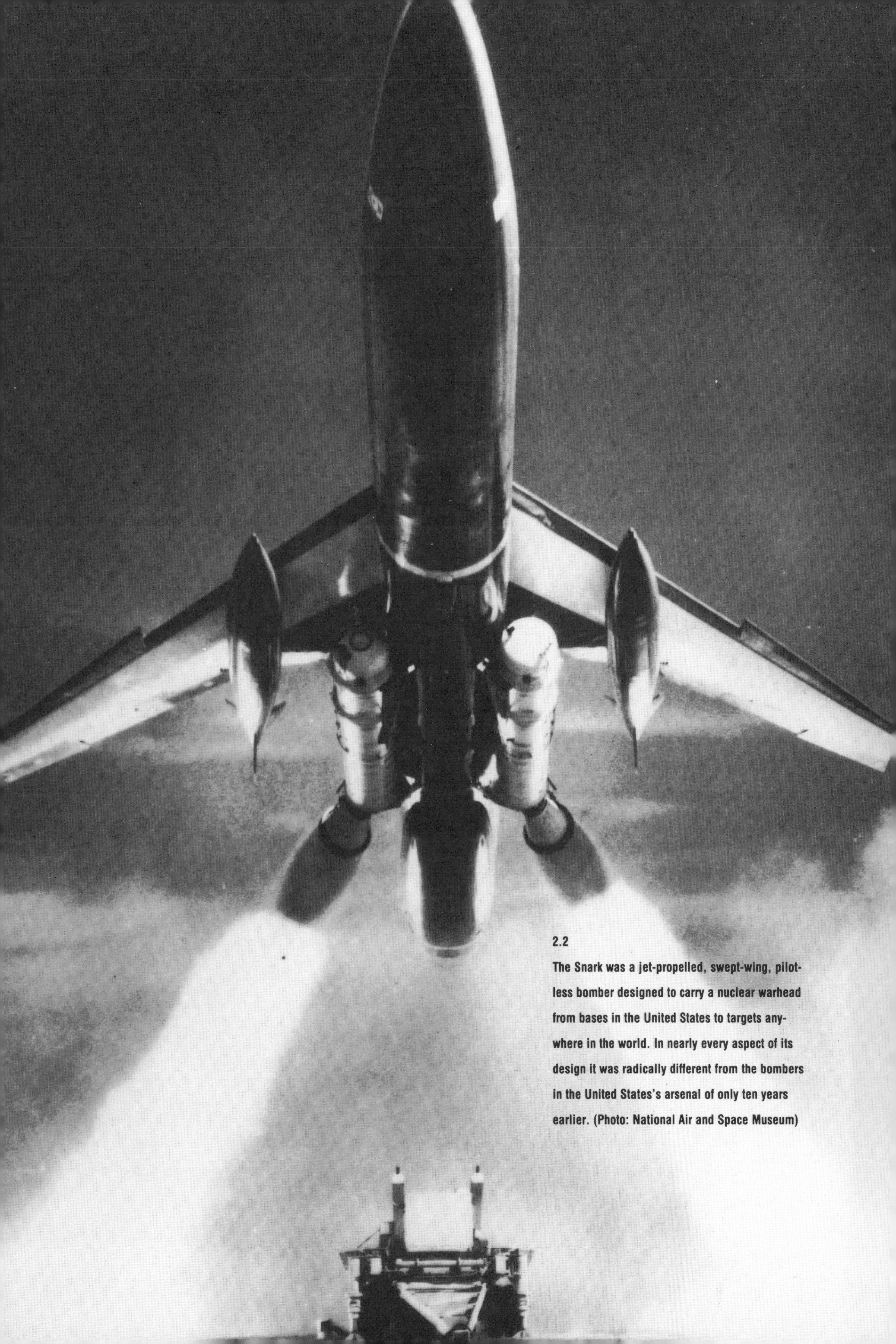

2.2

The Snark was a jet-propelled, swept-wing, pilotless bomber designed to carry a nuclear warhead from bases in the United States to targets anywhere in the world. In nearly every aspect of its design it was radically different from the bombers in the United States's arsenal of only ten years earlier. (Photo: National Air and Space Museum)

The principle of a celestial guidance system is simple: A system of telescopes and photocells senses the image and location of a star whose position in the sky is known.[2] The system compares the angle of the position of the star with an onboard reference point kept stable with gyroscopes and determines whether the missile is pointed in the right direction. It then issues commands to the missile's controls and engine to keep it on the desired course.

In practice, getting the system to do this without human intervention requires some method of translating information about the position of the star into a knowledge of the position of the missile, comparing that with the desired position, and automatically computing what commands would make up the difference if they do not agree. Guiding the Snark needed a computer, one that would solve the differential equations of motion that arose.

In 1946, when Northrop engineers were grappling with a guidance system for the Snark, they heard news of ENIAC, the computer built at the University of Pennsylvania for ballistics calculations. Those ballistics equations were similar to those required for describing the motion of the Snark. The next year Northrop hired John Mauchly, who with J. Presper Eckert invented ENIAC, to consult on the feasibility of a similar machine for guiding the Snark. Discussions with Mauchly led to a contract between Northrop and Mauchly's Electronic Control Corporation for an "experimental computer . . . to prove the feasibility of a particular method of navigation."[3] It was expected that the Electronic Control Company would ultimately provide a "compact, airborne" computer.

In retrospect it is hard to believe that the builders of ENIAC, which filled a room and consumed 174 kilowatts of power, were able to convince Northrop that they could provide a version of the same machine small enough to fly on the Snark. The news of ENIAC generated a lot of interest, and many descriptions of it emphasized the superiority of its use of vacuum tubes in digital circuits over mechanical analog techniques then common in aeronautic engineering. These advantages must have appeared compelling to those developing a guidance system as complex as that required for the Snark.

In 1949 Eckert and Mauchly delivered a machine they called BINAC (Binary Automatic Computer) as a ground-based "experimental computer" (figure 2.3). Like ENIAC it occupied almost an entire large room and consumed a great deal of power. They never delivered an airborne computer at all, and most accounts of BINAC state that it never worked after it was shipped from Philadelphia to the West Coast.[4] For Northrop BINAC was a failure, but Eckert and Mauchly received enough money from this contract to keep their company alive long enough to produce UNIVAC, of which they sold several. UNIVAC inaugurated the commercial computer business in the United States. Northrop thus unwittingly played a key role in moving the computer out of the laboratory and into commercial service.

MADDIDA

When it became clear that BINAC was of little use to the Snark project, a group of Northrop engineers, including Donald Eckdahl, Richard Sprague, and Floyd Steele, resumed work on a much simpler machine small and rugged enough to fly. Late in 1949 Northrop hired the Hewlett-Packard Company to build the first version of their design, which they called MADDIDA (Magnetic Drum Digital Differential Analyzer). As the name implies, it solved differential equations of motion by storing successive solutions of the equation on a magnetic drum about eight inches in diameter. Like BINAC it was a digital machine that used vacuum tubes for high speed, but unlike BINAC it was not a general-purpose computer.[5] Its design was a hybrid: it had the digital circuits of the new computers arranged like the analog computers of the earlier, mechanical era. The MADDIDA was compact and reliable (figure 2.4). It was still too large to be installed on board, but unlike general-purpose computers it contained all the design features needed to build a rugged and miniature version. Digital differential analyzers were, in fact, the first digital computers used for on-board missile guidance (for the Polaris). Yet when the Snark finally became operational in 1959, its guidance computer was not digital but analog.[6]

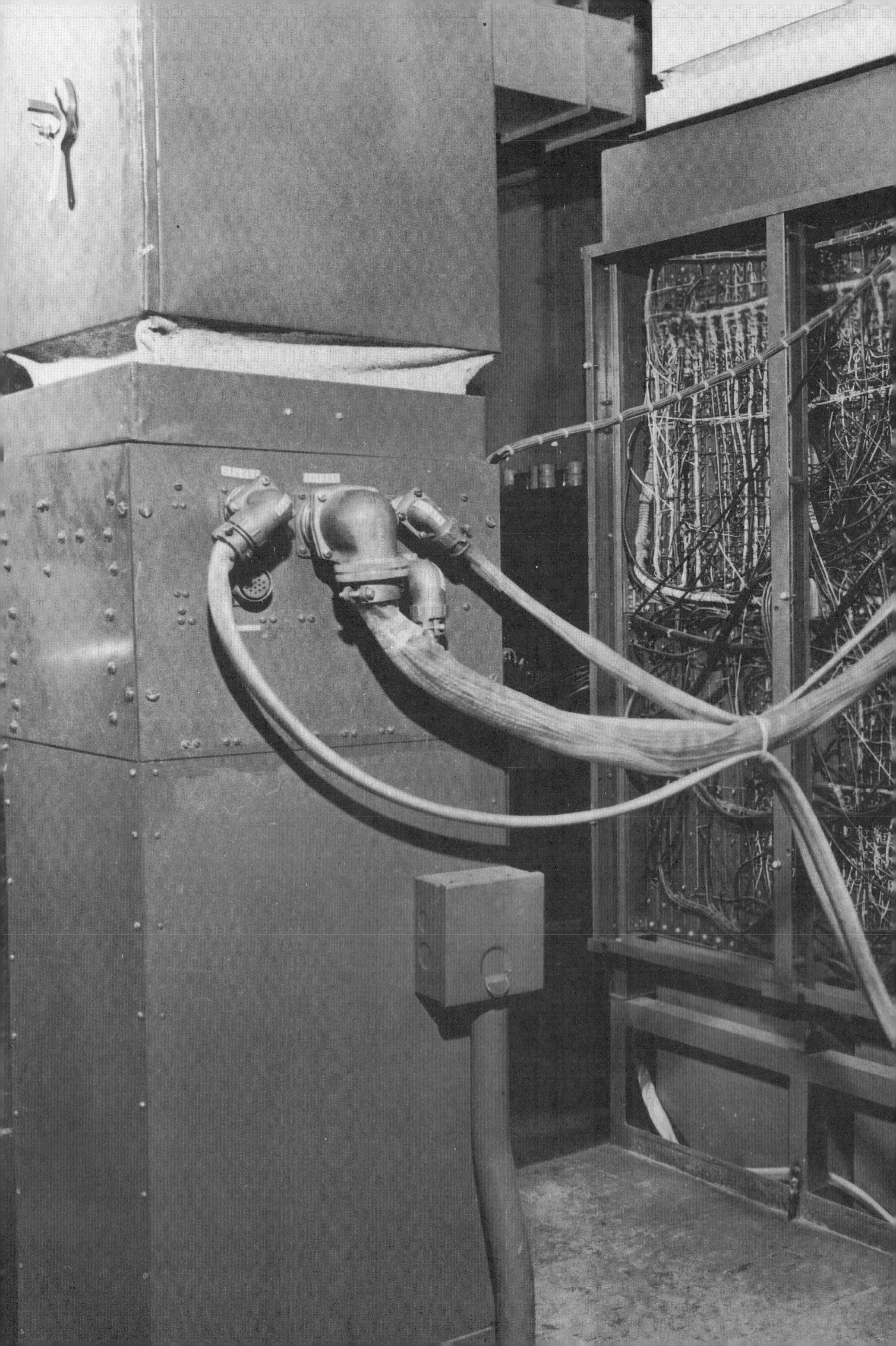

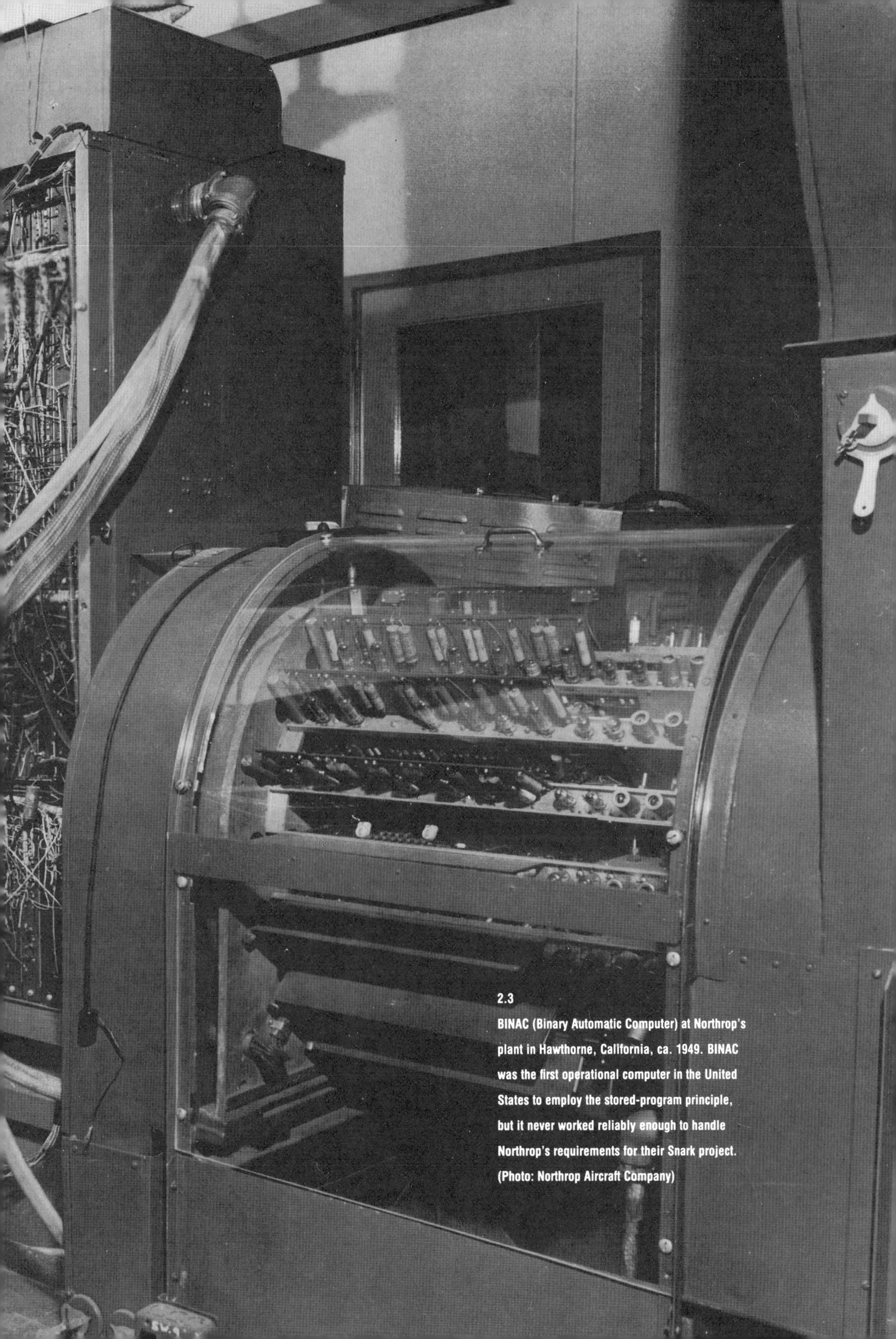

2.3
BINAC (Binary Automatic Computer) at Northrop's plant in Hawthorne, California, ca. 1949. BINAC was the first operational computer in the United States to employ the stored-program principle, but it never worked reliably enough to handle Northrop's requirements for their Snark project. (Photo: Northrop Aircraft Company)

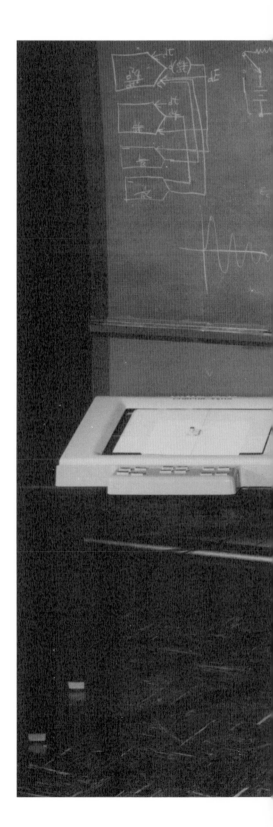

2.4
MADDIDA (Magnetic Drum Digital Differential
Analyzer) solved the differential equations of
flight by storing approximations to a solution on a
revolving drum. It was originally intended, but
never used, as an on-board computer for the
Snark, yet it became a successful commercial
product for Northrop. (Photo: National Air and
Space Museum)

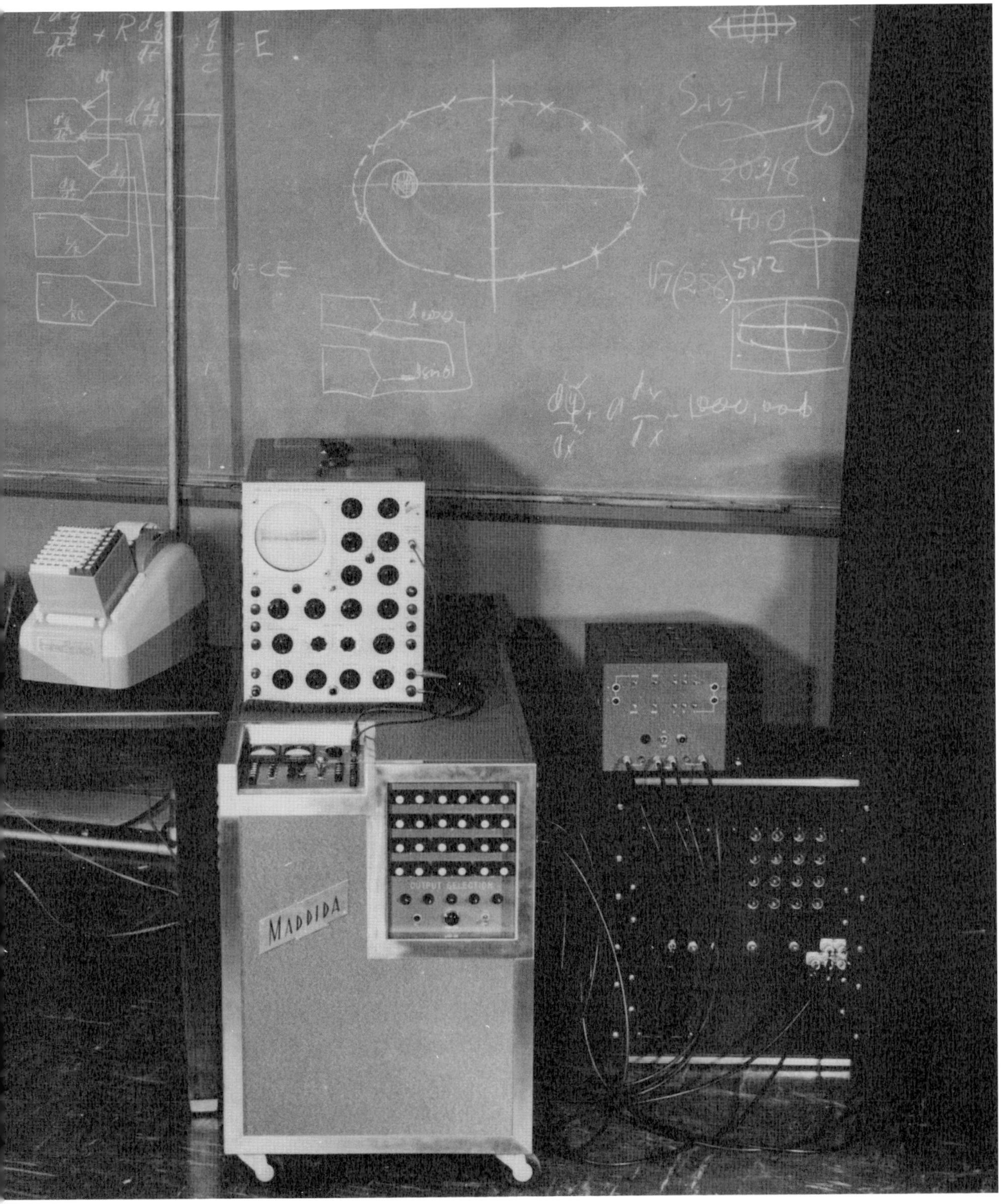
MADDIDA
OUTPUT SELECTION

The designers of MADDIDA were proud of their ability to build a computer that outperformed Eckert and Mauchly's in what they regarded as its most important respects. For them the fact that MADDIDA lacked a capacity for internally stored programs was more than offset by its ruggedness, compactness, and low cost. They suggested to Jack Northrop that the company open a separate division devoted to the sale of such computers to other aircraft manufacturers. But Northrop Aircraft's founder and namesake said no. (Northrop had previously considered selling computer time on BINAC, but the poor condition of that machine prevented them from doing so.) Development of both the Snark and the Flying Wing was not going smoothly, and the company was in financial trouble. In May 1950 about a dozen employees left Northrop and formed their own computer company, Computer Research Corporation (CRC). Their first product was a Digital Differential Analyzer, which they sold in respectable numbers, mainly to Southern California aircraft companies. With their next product, the CRC-102, they entered the general-purpose computer field. CRC then set a pattern that has since become the norm in the computer business. CRC grew rapidly on a base of technically superior products, spawned a number of other companies as its employees broke away from it, and eventually was acquired by an established firm (National Cash Register) seeking to enter the computing field. One study has traced the origins of no fewer than fourteen computer companies from the original Northrop group.[7]

The Card Programmed Calculator

MADDIDA was not the only contribution Northrop Aircraft made to computing. The exotic nature of engineering work on guidance and aerodynamics obscures the fact that in any aircraft company most of the engineers devote their time to the mundane and tedious work of structural analysis and testing.

An air or space craft must be strong enough to withstand whatever forces one may reasonably expect it to encounter during routine operations. To the initial calculation of the stress a craft encounters in steady flight in calm air, structural engineers add a load factor to ensure that it is strong enough to fly through turbulent weather or make sharp turns.

This load factor may vary according to the mission of the craft: a fighter, whose mission demands acrobatic and violent maneuvers, might have a much higher load factor than a passenger airplane. To the load factor the designer adds a factor of safety, thereby making the structure that much stronger. This factor does not counter unusual stresses that the craft is expected to encounter; it is part of the initial design estimates. The factor of safety ensures that the structure will hold up under the maximum stress it *may* encounter. Much of the factor of safety is there to compensate for what designers do not know about the strength of materials and structures, as well as for what they do not know about the conditions under which the craft will fly. As much as anything else, it is a factor of ignorance.

In a land vehicle a large amount of overbuilding is acceptable; indeed, in vehicles such as railroad locomotives extra weight has some advantages. But the added weight required to strengthen any structure limits airplane designers from applying this practice with abandon. Every additional pound of weight extracts a penalty in performance. Worse, it has a multiplier effect, as that pound requires more engine power and fuel to lift it, which in turn means more weight, which in turn requires a stronger and thus heavier structure to carry the extra fuel, and so on. There is thus ample incentive to reduce the weight of the airframe. Here is where computation comes in, and here in the structure division of aircraft companies, computing machines were most welcome.

In the early years of aviation such calculations were rarely done at all. The result was that airplanes with structures too weak crashed, while others flew safely but poorly because they were overweight. By the 1930s engineers had developed an adequate theory of structural analysis, which meant that the process of aircraft design could proceed on a more rational basis. But applying this theory to aircraft structures required a prodigious number of calculations, which in the 1930s and 1940s meant using mechanical calculators and people. The large rooms devoted to this work came to be known as "bull pens," because the low partitions that often separated the workers suggested a stockyard (figure 2.5).

2.5

In the years after World War II designing an aircraft was a labor-intensive process involving scores of engineers, drafters, and human "computers." (Photo: McDonnell-Douglas Corporation)

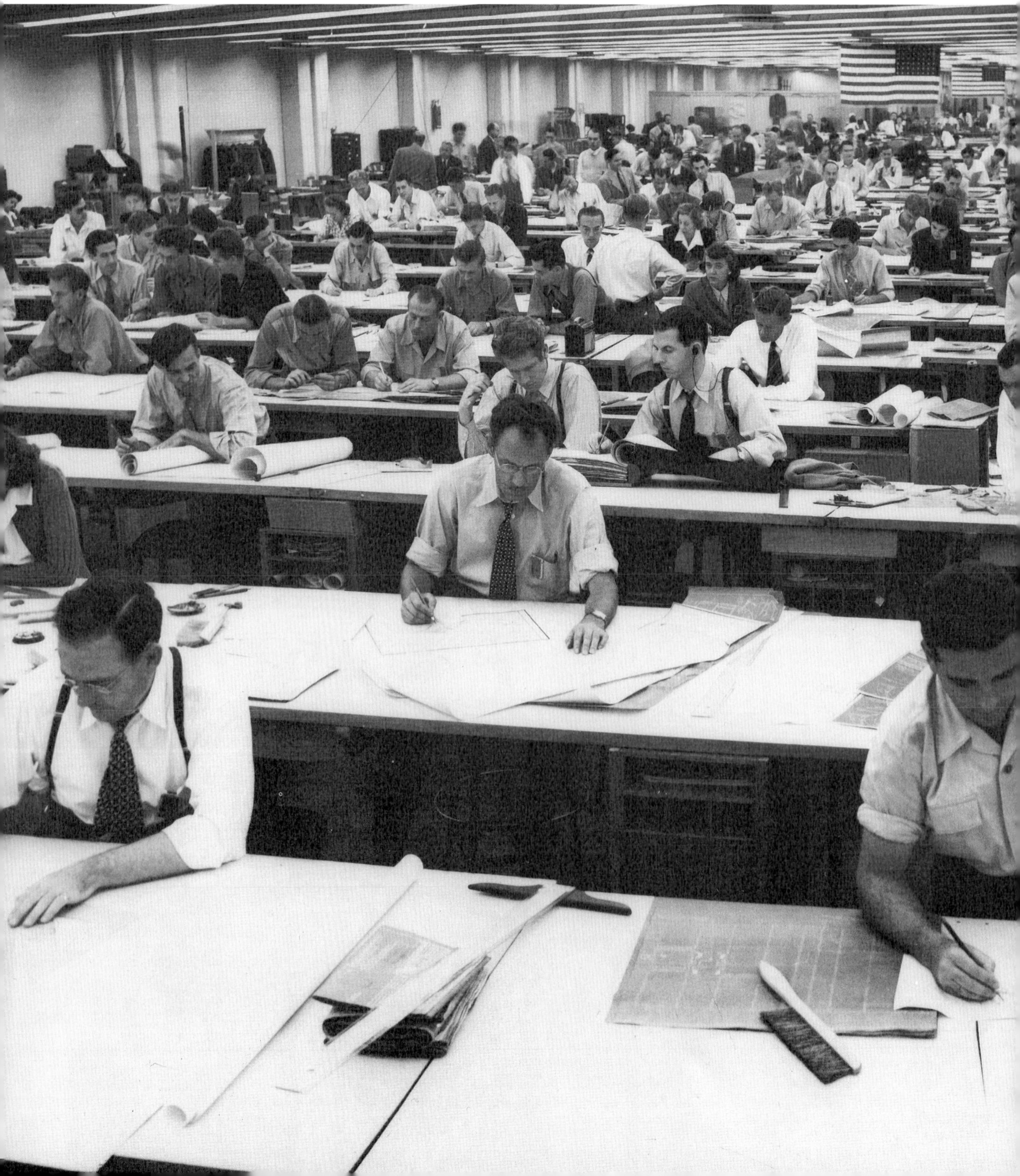

The most urgent problem was flutter. Almost anything—a flag, a leaf, a tall tree—will flutter and wave in a steady breeze. Airplane wings flutter too. If this flutter is not checked, the forces build up to greater and greater intensity, with catastrophic results (figure 2.6). Once begun, flutter can actually tear the wings off an airplane in seconds. As aircraft speeds increased during the 1930s, the flutter problem became more acute. Engineers had a theory they could turn to, which told them how to design wings with sufficient stiffness to ensure that they would not flutter, but solving the resulting equations required months of time and dozens of people. (This problem led Konrad Zuse to develop a mechanical digital computer in 1941 for the Henschel Aircraft Company of Berlin, as mentioned in chapter 1).

Problems like these pressed hard on aircraft companies. They required time and labor. A typical company would have most of its engineers working on structures, with even a separate division devoted solely to flutter analysis, and the flutter division would employ more people than the more glamorous aerodynamics division. One Northrop engineer described the scene as follows: "Try to imagine a room full of engineers, designers, and computation personnel, stretching as far as the eye could see, all operating adding machines, desk calculators, and comptometers. That was the aerospace computing center in 1945."[8]

In 1948 engineers at Northrop devised an alternate to this approach, one which involved them with the International Business Machines Corporation. Today IBM is synonymous with computers of all types, but in the 1940s their principal line of business was equipment that handled accounting, payroll, and statistical calculations for business. (It did serve and support a few scientific computing projects, but for IBM as for the rest of the calculating-machine industry, the scientific market was very small.) IBM machines operated on punched cards and handled numbers of modest size, seldom to more than the two decimal places required for calculations of dollars and cents. In 1946 IBM developed a vacuum-tube calculator that could multiply two numbers together and punch the results on a card in a few seconds. Compared to a mechanical desk calculator, this machine was bulky and expensive. But it

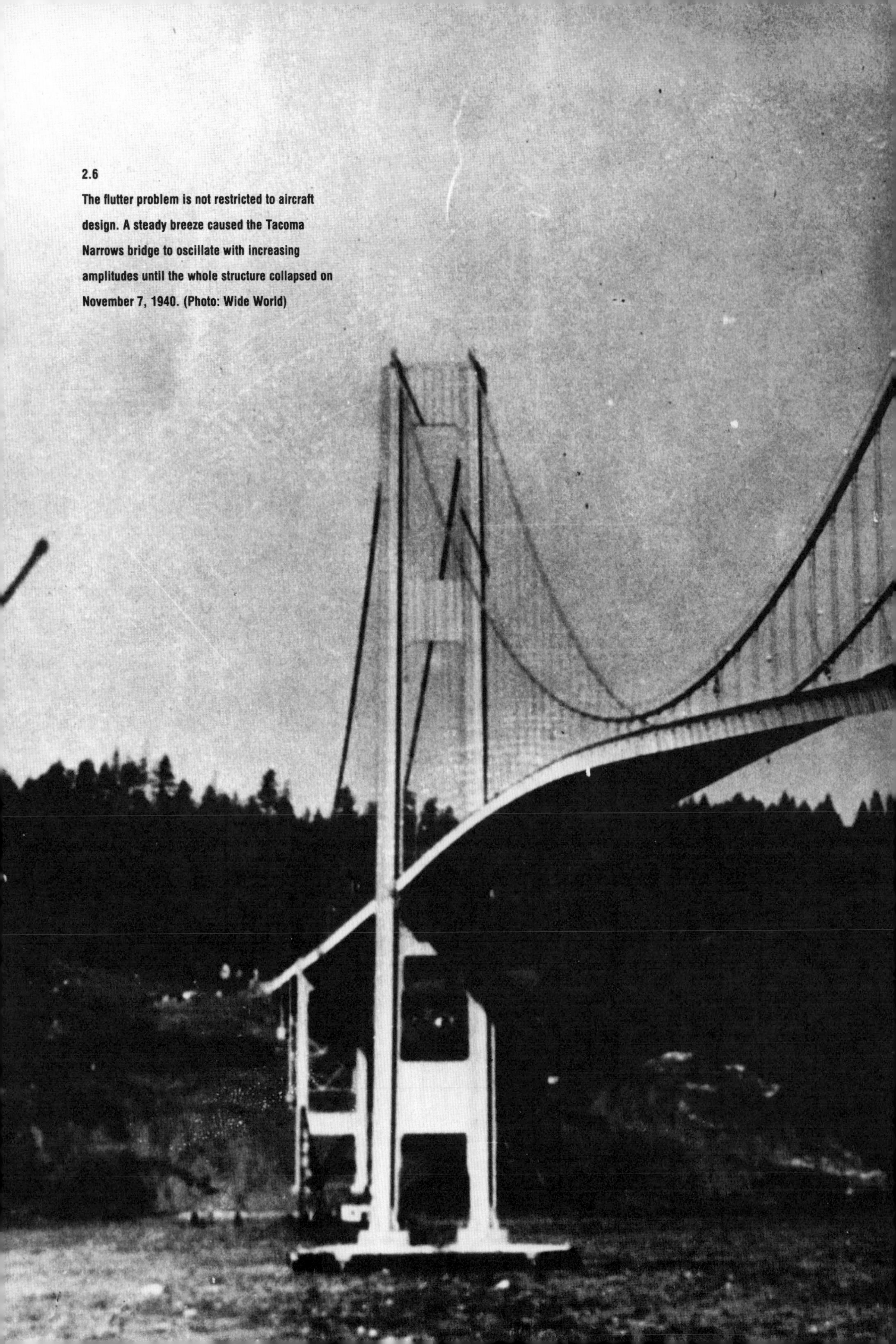

2.6
The flutter problem is not restricted to aircraft design. A steady breeze caused the Tacoma Narrows bridge to oscillate with increasing amplitudes until the whole structure collapsed on November 7, 1940. (Photo: Wide World)

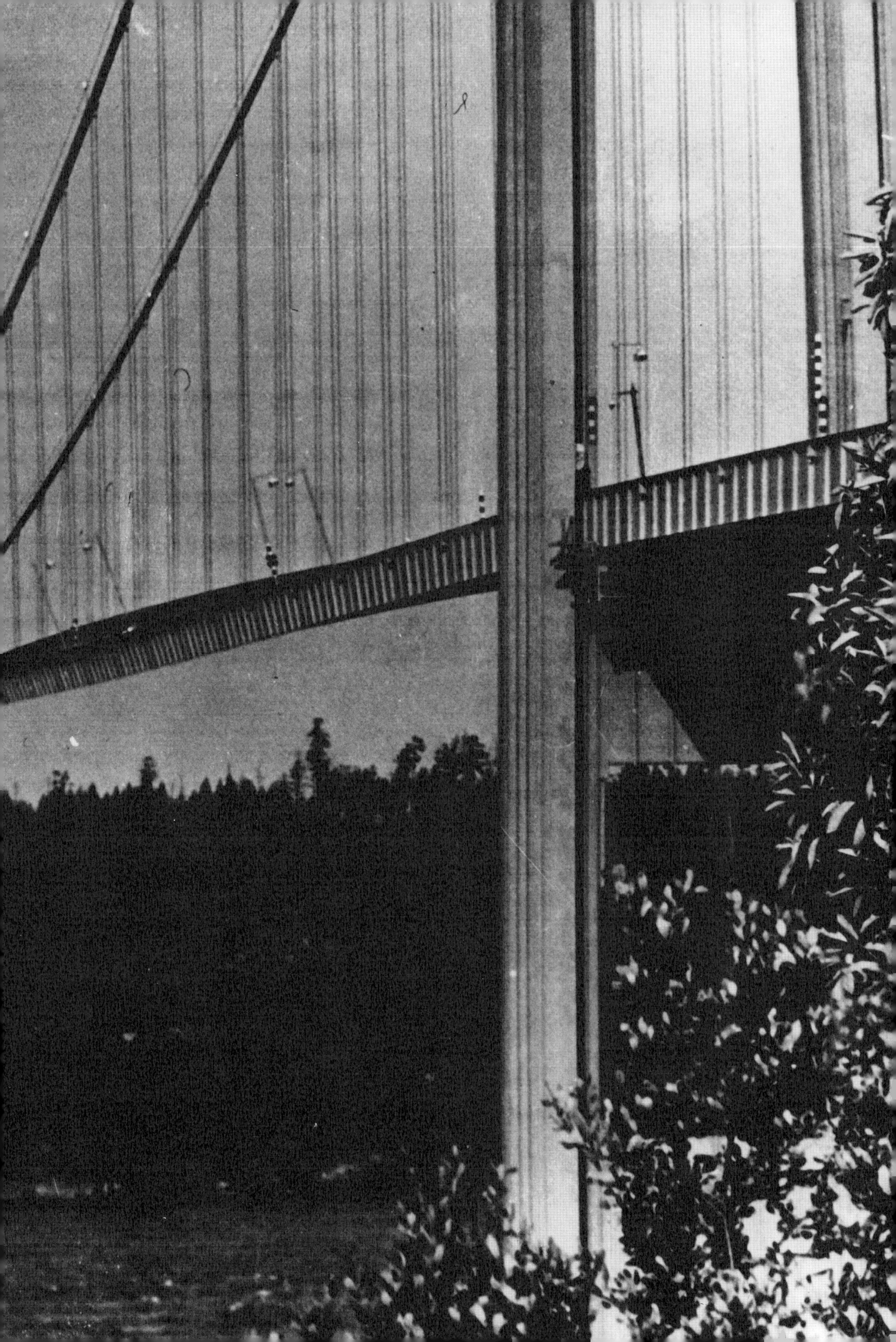

could work much faster. Moreover, it had a plugboard, patterned after a telephone switchboard. This gave it the ability to perform a short sequence of arithmetic operations on numbers punched into cards. An operator could change the sequence for each new problem by changing the plugs on the board. This latter capability made the IBM Model 603 calculator far more powerful than the desk machines then in use.

Northrop engineers, accustomed to spending weeks on a single problem using mechanical calculators, envied the accounting division of their company, which used these much faster IBM machines. So they "borrowed" a Model 603, connected it to another IBM machine that could print results on paper, and began using it for a variety of applications. IBM was at first surprised at Northrop's initiative. IBM's policy had always been to rent, not sell, its equipment, and this policy implied that the customer was not free to modify a machine at whim. But the success of Northrop's experiment convinced IBM that such a combination might be the basis for a new product they could sell to engineering firms that might otherwise not buy from IBM. In 1949 IBM introduced the Card Programmed Calculator, a commercial version of Northrop's experimental combination. It soon found a ready market among aerospace firms.

The Card Programmed Calculator performed arithmetic at high speeds using electronic circuits, but it did not store programs internally, as BINAC did. Programming the Card Programmed Calculator to solve different problems meant preparing a new plugboard and punching special cards that specified program steps. Typically an engineer would prepare a plugboard, punch the program cards, interleave the program cards with other cards on which numbers were punched, and run the deck through the machine.

A typical calculation that the Card Programmed Calculator performed was parameter variation, solving the same equation over and over, varying the input data a little each time. One example is computing how much runway an airplane needs to take off at a certain temperature, humidity, and barometric pressure. If this checked out all right, the problem would be solved for a slightly higher temperature, then a different humidity, and so on. This procedure would eventually produce a comprehensive table to which a pilot could refer during routine operations. For each variation of a parameter the design engineer would submit the same deck of cards over and over with only one or two data cards different.[9]

Unlike stored-program computers, the Card Programmed Calculator could carry out only short and inflexible sequences of arithmetic operations. But it was far superior to anything else available at the time. Based on standard accounting machinery that IBM had already installed elsewhere, the Card Programmed Calculator was dependable. For aerospace engineers this single fact outweighed any disadvantage it may have had compared to the electronic computers just then being developed.[10] In the immediate postwar years, during which aerospace saw a great transformation, the Card Programmed Calculator, punched card equipment, mechanical calculators, and special-purpose analog and digital computers like MADDIDA were the chief computing tools that enabled aerospace firms to enter the jet age and space age.

Chapter 3
Logistics and the RAND Corporation

For a picture of what life was like for an ordinary soldier during World War II, there is no better place to turn than to Bill Mauldin's cartoon character Willie (figure 3.1). Willie always needed sleep, a shave, and a meal more palatable than K rations. He spent most of his time waiting for orders telling him to go somewhere, and when he got there (on foot), he had to wait some more to find out what to do next. Behind the humor is a serious problem that military planners ever since Napoleon have tried to solve scientifically. The problem is logistics: getting the right material to the right men at the right time in the right place. For the air branches of the U.S. Army and Navy, it meant primarily getting enough airplanes to the right places, but that was only the first step. It meant also getting fuel, oil, and tires to those airplanes, maintaining them, building landing strips for them, and last but not least, training men and women to fly them.

3.1

Bill Mauldin volunteered for the Army a year before Pearl Harbor and spent three years in the infantry. He recorded his observations about Army life in a series of Pulitzer-Prize winning cartoons that first appeared in the Army newspaper *Stars and Stripes*. (Bill Mauldin, United Feature Syndicate)

"Git his pistol, Joe. I know where we kin swap it fer a combat jacket an' some boots."

During World War II clerks "armed" with pencil and paper did this job. To provide room for these men and women the War Department in 1942 built the Pentagon, for years the largest office building in the world. Toward the end of the war the Army Air Forces turned to IBM and Remington Rand for accounting machines they built and sold to private corporations to handle inventory and purchasing. These machines and their stacks of punched cards did the job, but just barely.

SCOOP

One of the first projects the Air Force began after its founding was Project SCOOP, an acronym for "Scientific Computation of Optimum Problems." Its mission was to find a scientific method to manage the logistic problem of operating an air force composed of jet fighters and bombers, ordinary fighters and bombers, and ballistic and guided missiles, all at air bases scattered around the world. In a typical case they wanted to train a maximum number of pilots in a minimum time period with a minimum number of training airplanes and instructors. One of the first employees to work on SCOOP was the mathematician George Dantzig, who had done similar work in civilian economics modeling.

The mathematical expression of these problems was typically a set of linear inequalities (e.g., the number of pilots is over 1,000; the time for training them is less than 90 days; etc.). Dantzig drew on his experience to reformulate the problem in terms of equations. Each equation might have many different variables representing unknown quantities, but no variable was raised to a higher power. The graph of a such an equation forms a straight line (by contrast, an equation having squared quantities forms a parabola or an ellipse); hence, mathematicians called them "linear" equations, and the process of solving a set of linear equations "linear programming."[1]

By 1948 Dantzig developed a technique for solving these sets of equations and proved that such a solution would be the optimal one for the original problem. The method he developed, now known as the "simplex" method, has since become a cornerstone of modern applied mathematics.[2]

The sheer number of equations needed to describe the logistics problems Dantzig faced implied an enormous amount of computation. To test the technique, he sought to develop the least expensive diet that would meet the average adult's minimum daily requirements for nine essential nutrients. It was a problem well known to Dantzig's colleagues in applied mathematics. Its solution by the simplex method required about seventeen thousand multiplications and divisions, and took clerks using mechanical desk calculators twenty-one working days. The test demonstrated that the simplex method worked. But the time required made it clear that only electronic computers, not human beings, could solve the much more complex problems the Air Force needed to solve.

In 1948 electronic computers were not easy to come by. Dantzig explored using BINAC to solve a similar test problem, but BINAC, as we have seen, was not operational. He developed techniques for solving the problem on punched-card business machines and ran a test in which he fed four thousand cards into a set of IBM punched-card multipliers, which then ran for the next eight hours, all for the solution of a deliberately simplified problem.[3]

Though it wanted to, the Air Force could not just go out and buy a computer that would fill its needs. When in 1946 the National Bureau of Standards issued a call for commercial bids to build a computer for the 1950 census, only two companies came forward. One was Eckert and Mauchly's Electronic Control Corporation, already mentioned, and the other was Raytheon, a Massachusetts electronics company that had worked on radar during the war. The National Bureau of Standards judged that Raytheon's bid was too high. Eckert and Mauchly's company bid less, but the company was so small and strained for resources that it was obvious they could not deliver a working machine in any reasonable time. The small company's difficulties with the BINAC contract for Northrop revealed their inability to deliver finished commercial products, which was what the Air Force wanted. The Bureau of Standards did award Eckert and Mauchly a contract for the census computer, UNIVAC, which was eventually completed in 1951 (too late for the 1950 census) and not delivered to the Census Bureau until 1952. (Electronic Control Corporation was soon absorbed by Remington Rand.)

The Air Force then contracted with the National Bureau of Standards to quickly build a computer to serve as an Interim machine until UNIVAC was available. The Interim computer was completed in May 1950 and was immediately pressed into service to explore Dantzig's simplex method.[4] The Bureau of Standards built the computer at their facilities in Washington, D.C.; a few years later they built another for a branch of the bureau in Los Angeles. The Interim computer was rechristened SEAC, for "Standards Eastern Automatic Computer," and had a long and productive life, serving many customers besides the Air Force. SEAC worked very well. It was the first operational stored-program computer in America.[5]

The linear programming problems that arose in Project SCOOP required input and output facilities that could handle large volumes of data at high speed. SEAC lacked this ability, at least in its early configuration, and so was not as useful for logistics problems as the Air Force had hoped. Yet it was useful for other problems and served not only as a workhorse computer but at the same time as a test-bed for experimental devices designed to improve its memory and input-output capacity. In June 1952 the Air Force installed a UNIVAC in the Pentagon, and with this machine they finally had a practical means of realizing Dantzig's theoretical innovation. This UNIVAC was only the second one built by Eckert and Mauchly. It was the first of their stored-program computers that was shipped from their factory, was installed in a customer's site, and performed useful work there. This UNIVAC (serial no. 2) helped inaugurate the commercial computer industry in the United States.[6]

Today a computer with the power of UNIVAC can fit into a box about the size of a sewing machine, and it is just as easy to move. But moving UNIVAC to the Pentagon was not so easy. Well ahead of the installation day, engineers met with military personnel to ensure that the giant pieces of UNIVAC would fit through Pentagon doorways, and that the building's floors would support the thirty thousand pounds that the computer weighed. UNIVAC required a lot of power to run. Since most of this power went to heat the vacuum tubes, it needed an elaborate cooling system to carry this heat away. It took three months to complete the move. Once in the Pentagon, security restrictions required the Air Force to take on the job of operating and maintaining the computer, not an easy task for something that complex and novel and a job that the Census Bureau was not capable of doing.[7]

RAND

The intellectual focus of Project SCOOP was not at the Pentagon but in Southern California, specifically at an unusual place called the RAND Corporation. The Army Air Force conceived of this organization in 1945 as a way to keep the expertise of civilian scientists, so crucial to the war just ended, from slipping away. It began with a $10 million appropriation to Douglas Aircraft to house and administer an agency that would employ researchers for projects deliberately left unspecified at the time. RAND immediately took on some challenging problems relating to aircraft and rocket propulsion systems, waging nuclear war, and the logistics problems discussed above. RAND had a strong concentration of talent in applied mathematics, and in 1953 it built a digital computer of its own. The computer was called Johnniac, a witty comment on the genius of "Johnny" von Neumann, whose ideas on computer design had a great impact on the technology after 1945. Von Neumann is said to have been embarrassed by this name.[8]

The Air Force gave RAND a free hand in pursuing its research. At least that is what nearly every one of its employees in its computational division recalled about the place when interviewed twenty years later.[9] Some of the work they did bears this out. (The acronym "RAND" officially stood for "Research and Development," but some have thought it meant "Research and No Development.") At a time when even getting a computer to work at all for more than a few minutes was an heroic engineering achievement, RAND employees were writing reports on using computers to "think" like human beings. Their reports had little bearing on immediate Air Force needs, but they laid the foundation for nearly all of what is now a mainstream activity of computing, artificial intelligence.[10] Another group at RAND developed a method of programming Johnniac so that one did not have to be a computer expert to use it. In short, RAND was a center for fundamental research in what is now called computer science a good decade before either the subject or even the term became known and understood as one worthy of serious study.

By 1948 RAND had become an independent, nonprofit corporation and had severed its connection with Douglas Aircraft. It was not completely independent, as it still received all its work in the form of contracted research from the U.S. Air Force. But just as it was free to pursue a wide range of intellectual ideas, so it was free to work with computers in ways not tied to a specific aircraft company or Air Force project. At this time all of the Southern California aircraft companies were struggling to get the first commercial computers working. In 1953 the IBM Corporation began shipments of a stored-program computer on a par with Remington Rand's UNIVAC. This machine, which they called Type 701, had a processing speed, storage capacity, and program capacity that far surpassed its Card Programmed Calculator or other punched card equipment. It was an expensive computer, renting for about $15,000 a month. But by 1955 most Southern California aerospace companies had acquired one. Of the total nineteen Type 701 computers IBM built, eleven were for aerospace customers, seven of those in Southern California (table 3.1). And when in December 1955 IBM announced an improved version called the 704, these Southern California companies put in their orders for that one as well.[11]

In those days when one got a computer, it came with almost no instructions or guides for effective use. Programming meant preparing decks of cards with instructions in an obscure numeric code. Even to get the computer to extract a square root involved quite a lot of such work. Worse, the program codes required binary or octal numbers, which few aircraft engineers had ever seen before. It is an indication of the great need for computing power for aircraft design and manufacturing that as many companies obtained these expensive machines as did.

In this environment RAND was a clearinghouse and neutral ground where Douglas, Northrop, North American, and Lockheed could swap programming tips, maintenance procedures,

Table 3.1 Original 701 customers

Machine number	Destination	Date
1	IBM World Headquarters New York, N.Y.	12/20/52
2	University of California Los Alamos, N.M.	03/23/53
3	Lockheed Aircraft Company Glendale, Calif.	04/24/53
4	National Security Agency Washington, D.C.	04/28/53
5	Douglas Aircraft Company Santa Monica, Calif.	05/20/53
6	General Electric Company Lockland, Ohio	05/27/53
7	Convair Fort Worth, Tex.	07/22/53
8	U.S. Navy Inyokern, Calif.	08/27/53
9	United Aircraft East Hartford, Conn.	09/18/53
10	North American Aviation Santa Monica, Calif.	10/09/53
11	RAND Corporation Santa Monica, Calif.	10/30/53
12	Boeing Corporation Seattle, Wash.	11/20/53
13	University of California Los Alamos, N.M.	12/19/53
14	Douglas Aircraft Company El Segundo, Calif.	01/08/54
15	Naval Aviation Supply Philadelphia, Pa.	02/19/54
16	University of California Livermore, Calif.	04/09/54
17	General Motors Corporation Detroit, Mich.	04/23/54
18	Lockheed Aircraft Company Glendale, Calif.	06/30/54
19	U.S. Weather Bureau Washington, D.C.	02/28/55

Source: Manufacturing records, Poughkeepsie, N.Y. See C. C. Hurd, Edited Testimony, *Annals of the History of Computing*, 3 (April 1981), p. 169.

and the like without fear of giving away too much to their competitors. In 1955 RAND and about eighteen mostly Southern California aircraft companies founded SHARE, a users group led by Paul Armer, head of RAND's Numerical Analysis Department, for the purpose of sharing information on the IBM Type 704 computer they each were using at the time. Like RAND's other work in computing, SHARE was vital to the diffusion of computing technology into aerospace in the 1950s. This was the first of many such groups bound by use of a common machine. Its existence and work reveal the extent to which the aerospace industry had to be an active, if reluctant, part of the computer industry when the computer industry was in its infancy and the aerospace industry was being transformed.[12]

Chapter 4

Missile Tracking and the Search for Real Time

4.1

A theodolite is a telescope mounted on gimbals so that it is free to move along both the vertical and horizontal axes. In its more sophisticated form the telescope is connected to a radar unit that automatically tracks the target. A movie or television camera connected to the telescope's eyepiece gives a visual record of the event. One can also record the elevation and compass angles the telescope moves through to provide tracking data about where the object is in three-dimensional space. (Photo: Naval Air Test Center, Patuxent River, Maryland)

4.1

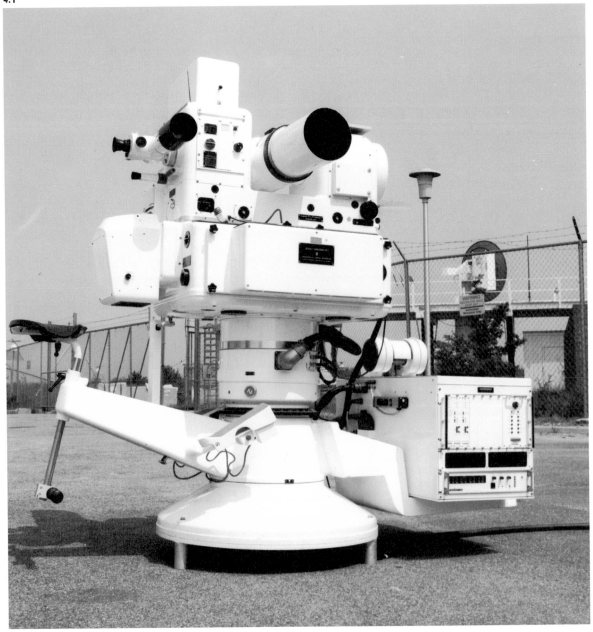

For someone familiar with the operation of aircraft as they existed in the late 1940s, learning to deal with ballistic missiles must have been a perplexing and frustrating experience. For one, there was no pilot. Also, the aerodynamics were different: ballistic missiles fly not *in* the air but rather *through* the air into the upper atmosphere and beyond and finally back through the air to a target. Finally, the flight path of a ballistic missile is for the most part not under anyone's control: only in the initial boost phase, when its engines are burning, can one do much about a missile's trajectory. After that, it travels in a free-fall arc just like any thrown object (hence the name "ballistic," from the name the Romans gave to a catapult, which hurled large rocks at an enemy fortification).

Despite these difficulties, using ballistic missiles to deliver bombs gained advocates among U.S. military strategists, including many at RAND, as the best way to carry a war to the Soviet Union. Missiles did not risk the lives of crewmen. Long-range missiles did not require expensive and politically delicate U.S. air bases in European or Asian countries. But if they were to work, there had to be a way to assure that the missile was on the proper course in those first seconds of its flight, when its engines were running and control was possible.

That requirement led to a number of technical problems, each needing an urgent solution. Someone has to track the missile at launch to know precisely where it is and how fast and in what direction it is traveling. One then has to compare that information with previously obtained tables of where the missile *ought* to be to hit a given target. And finally, one has to command the missile to alter its trajectory if the two sets of information do not match. On-board instruments can measure how the missile is performing; telemetry can then transmit this information to ground controllers. On the ground below the flight path, telescopes mounted on calibrated pivots can track a missile up to a range of forty miles, depending on visibility. These instruments are called theodolites and are the mainstay of all forms of airplane and missile tracking (figure 4.1). Radar sets on the ground can also measure the range and elevation of a missile. A technique called Doppler radar can pinpoint a missile's velocity accurately by measuring the change in frequency of a signal sent from the missile, just as one can tell the approach and passing of a train by the change in the pitch of its whistle.

The information these techniques provide was not in a form that controllers on the ground could use. Theodolites and radars give angles of elevation and compass directions; radar gives the distance of the rocket from the radar station. Ground controllers wanted to know where the missile was in terms of its altitude, its location over a point on the ground, and its velocity. Converting the range data to this form was an old problem of geometry (it is called resolving, and machines that did it were called resolvers). The basic calculation is a matter of taking the sines and cosines of the tracking angles and multiplying them by the range data. A mechanical analog computer can resolve simple problems (figure 4.2), but these were usually too crude for tracking missiles. This job was just the kind that electronic computers were good at performing.

An additional requirement cast the entire problem of tracking and guidance into a different mold. To be effective, a computer had to process the data as the missile was in flight, fast enough to transmit corrective actions before the rocket motors cut off. Firing-range engineers called this computing in real time. It meant a computing speed many times faster than that of even the fast electronic computers just invented.

During World War II the U.S. Navy established a branch called the Special Devices Division to exploit advances in electronics and other technologies primarily for Navy aviation. The Special Devices Division was the brainchild of Navy Captain Luis DeFlorez, a graduate of the Massachusetts Institute of Technology and a man who had distinguished himself during the war for his work on building flight simulators to train Navy aviators (figure 4.3). At the end of the war the Special Devices Division initiated a series of projects, each with a distinctive name such as Whirlwind, Cyclone, Typhoon, and Hurricane. The goal of Project Hurricane was to track and control ballistic missiles.

θ, degrees
90°
80°
70°
60°
50°
40°
30°
20°
10°
0°
1.0
0.8
0.6
0.4
Sine θ
0.2 0.4 0.6 0.8 1.0
Cosine θ

θ, degrees
90°
80°
70°
60°
50°
40°
30°
20°
10°
0°
1.0
0.8
0.6
0.4
Sine θ
0.2 0.4 0.6 0.8 1.0
Cosine θ

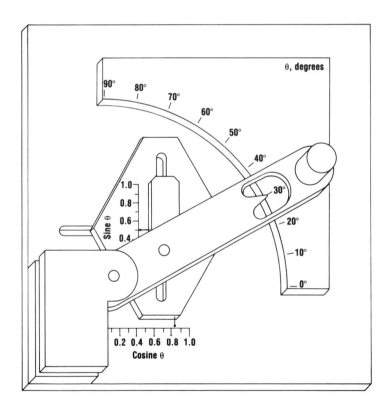

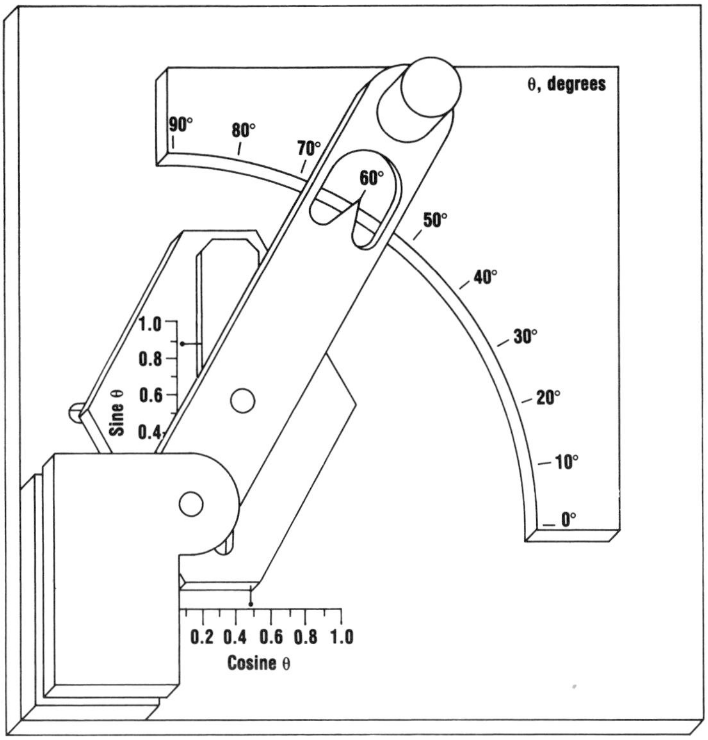

4.2

This simple mechanical arrangement of pins and levers computes the trigonometric functions of the sine and cosine of an angle. The computation is performed continuously and instantly, but its accuracy is limited by the precision of the machining of the parts, usually around one percent. In its most common aerospace application it is connected to a tracking theodolite and a range radar. Using angular data from the theodolite and range data from the radar, it computes the altitude above ground and distance downrange of an air or space craft.

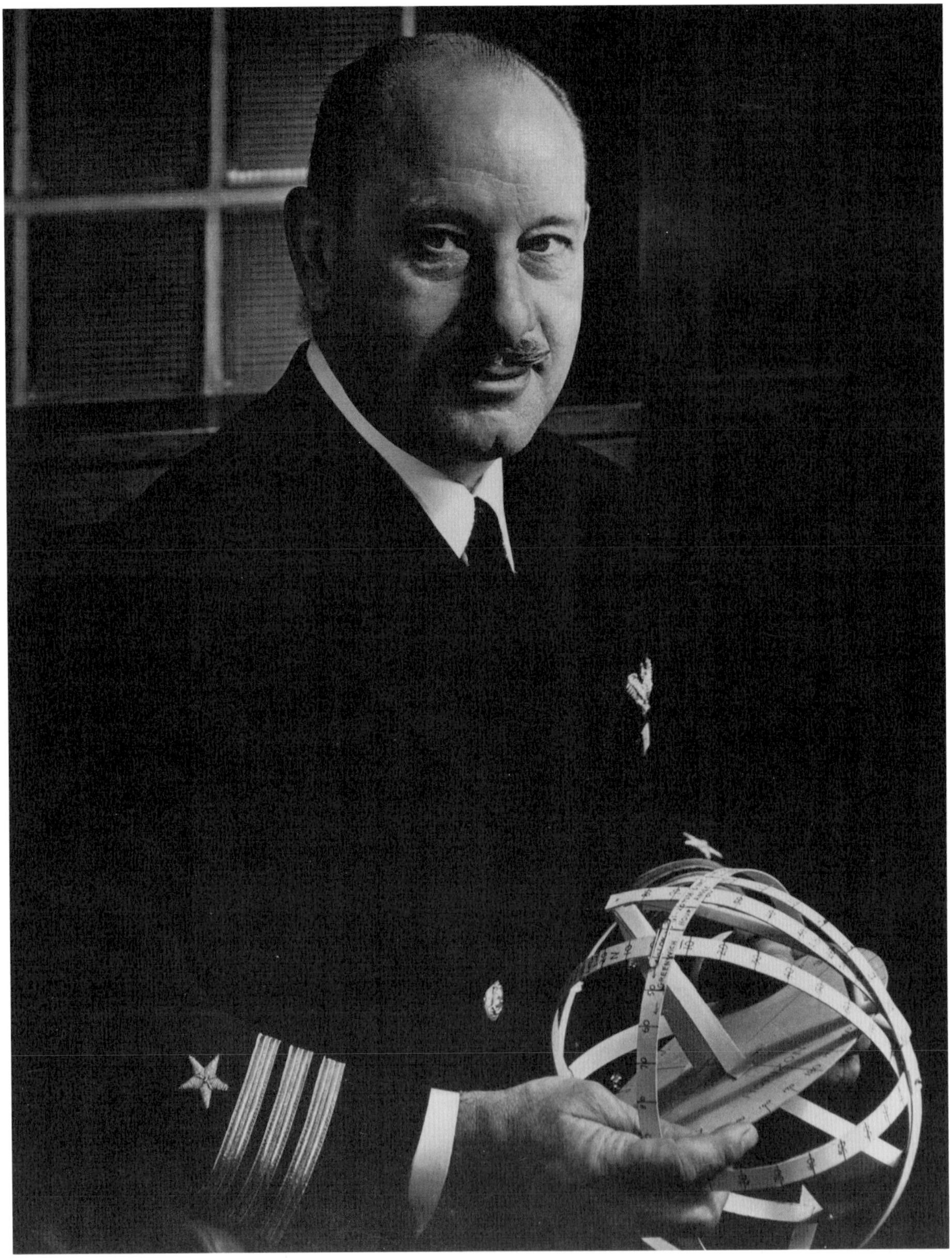

4.3

Project Hurricane was intended to provide tracking and guidance at the Naval Air Missile Test Center at Point Mugu, California, and at Holloman Air Force Base, New Mexico. It was a complex system of theodolites, Doppler and conventional radars, and telemetry systems. Its central nervous system was a computer. This computer had to be fast enough to act in real time, and it was to be a digital rather than analog computer—a combination never before achieved.[1]

Analog versus Digital

Consider once again the difference between analog and digital computing devices. The watches described in chapter 1 tell time by scaling the motion of an escapement or a quartz crystal so that it matches the duration of seconds, minutes, and hours. That is a simple calculation, but once it is set by the watch design, it does not vary. An analog device that does more complicated arithmetic is the differential gear found in the rear axle of a typical automobile (figure 4.4). When a car goes around a corner, the two rear wheels must turn at different speeds, though each continues to receive power from a single driveshaft. The differential gear distributes the engine's power so that the sum of the rotations of the two rear wheels is proportional to the rotation of the driveshaft. Note how that sum can vary depending on whether or not the car is traveling in a straight line. In other words, the differential gear computes a simple, continuous weighted average.

The action of these gears is unlike a digital adding machine in several important ways. First, this machine computes continuously and smoothly as fast as the input comes in. A digital adding machine by contrast requires some time for one to enter the two numbers to be added, followed by some time to do the addition. Only then does it produce the sum.

Second, the gear's output is a sum in the form of not numbers but rather torque, which is exactly what the automobile engineer wants here. The output could just as easily be to a numeric dial, but in this instance that is not what is required. This characteristic is not unique to analog mechanical com-

4.3

Shortly after the United States entered World War II in 1941, Navy Captain Luis DeFlorez established a division of the Navy devoted to training flight crews with simulators. Out of this effort came Projects Whirlwind, Hurricane, Typhoon, and Cyclone, each of which simulated an aspect of air or space flight. (Photo: National Air and Space Museum)

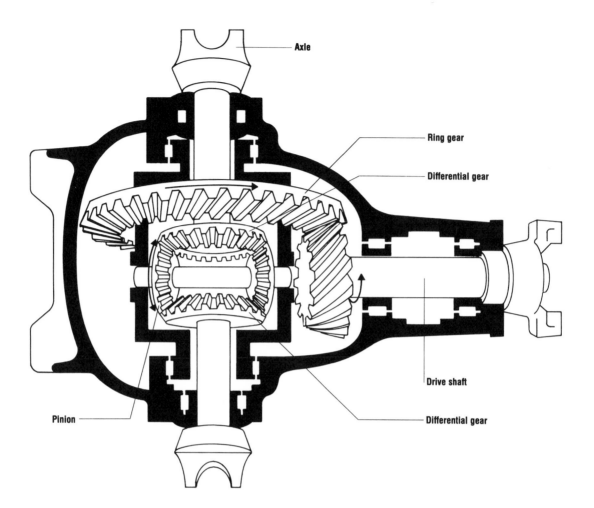

Axle

Ring gear

Differential gear

Drive shaft

Differential gear

Pinion

4.4

The familiar automobile differential gear is a
good example of the computation of a short mathematical expression by constructing a mechanical
analog. Here the analog uses gears.

puters, but it is typical. A digital adding machine typically displays its sum as a number. Its output can be converted into a mechanical motion, but in practice it is somewhat more complex to do so.

Finally, if one does not require a lot of force at the output, mechanical analog computers can be made very compact, rugged, and maintenance-free. That was not so with digital computers available in the 1950s.

Each of these factors favored analog over digital approaches to the design of computers carried on airplanes or rockets. Bombsights, navigational instruments, and autopilots all required the attributes that analog computers offered, especially continuous, or real-time, operation. By 1950 the development of mechanical analog computers for these functions had reached a high state of sophistication.

Why then the decision to use a digital computer for Project Hurricane? For one, the Hurricane computer was to be located on the ground, where size was less an issue. But the main reason was the need for flexibility. Analog mechanisms can do simple addition, trigonometry, and integration, but to combine more than a few of these functions requires complicated trains of gearboxes, driveshafts, and other connecting devices. As complexity increases, accuracy rapidly decreases. And solving a different equation requires tearing the whole machine apart and rebuilding it from scratch. These were the only drawbacks of analog devices, but for a project like Hurricane they were compelling. As digital technology advanced and overcame the other objections mentioned above, it displaced analog computers from their entrenched position throughout aerospace.

RAYDAC

The Hurricane computer had its origins in the National Bureau of Standards' earlier attempt to obtain a digital computer for other federal agencies including the Census Bureau and the Air Force. Recall that the National Bureau of Standards decided to build the Interim computer after failing to obtain an acceptable bid for such a machine from any commercial manufacturers. Besides Eckert and Mauchly's bid, the Raytheon Manufacturing Company of Waltham, Massachusetts, was the only other company to bid. The National Bureau of Standards rejected the bid, but the Navy later contracted with Raytheon to build this machine for Hurricane. In keeping with the tradition of using acronyms, Raytheon christened this computer RAYDAC, for "Raytheon Digital Automatic Computer" (figure 4.5).

RAYDAC had a number of advanced design features. It used a sophisticated method of storing data on magnetic tape. Its designers also pioneered a method of error checking that ensured that the machine would detect any fault in its circuits and halt operations before allowing an erroneous number to be transmitted to another part of the computer. (The technique, called parity checking, is now standard in all modern computers.) But RAYDAC was simply not fast enough for the real-time operation requirements of Project Hurricane. Its memory unit stored digits in the form of sound pulses that circulated round and round in a tank of mercury, an idea borrowed from radar technology. To retrieve a piece of information, the computer had to wait until the acoustic pulses came around to a detector where they could be read. It was much like trying to read a news bulletin from a single-column ticker-tape. This waiting time might be only a few thousanths of a second, but it was still too long for real-time missile guidance and tracking.

RAYDAC was a failure for Project Hurricane. But I must qualify that statement. The computer worked as well as or better than most of its kin elsewhere. It handled other voluminous calculations needed at Point Mugu. It advanced the state of the art of both computer technology and missile tracking. It helped tip the balance away from manned bombers toward long-range ballistic missiles as the primary strategic weapon of the United States. And for the computer industry, it brought into sharp focus the technical hurdles that real-time, digital computing had to overcome. And they were overcome, largely by the next project we shall turn to, Whirlwind.

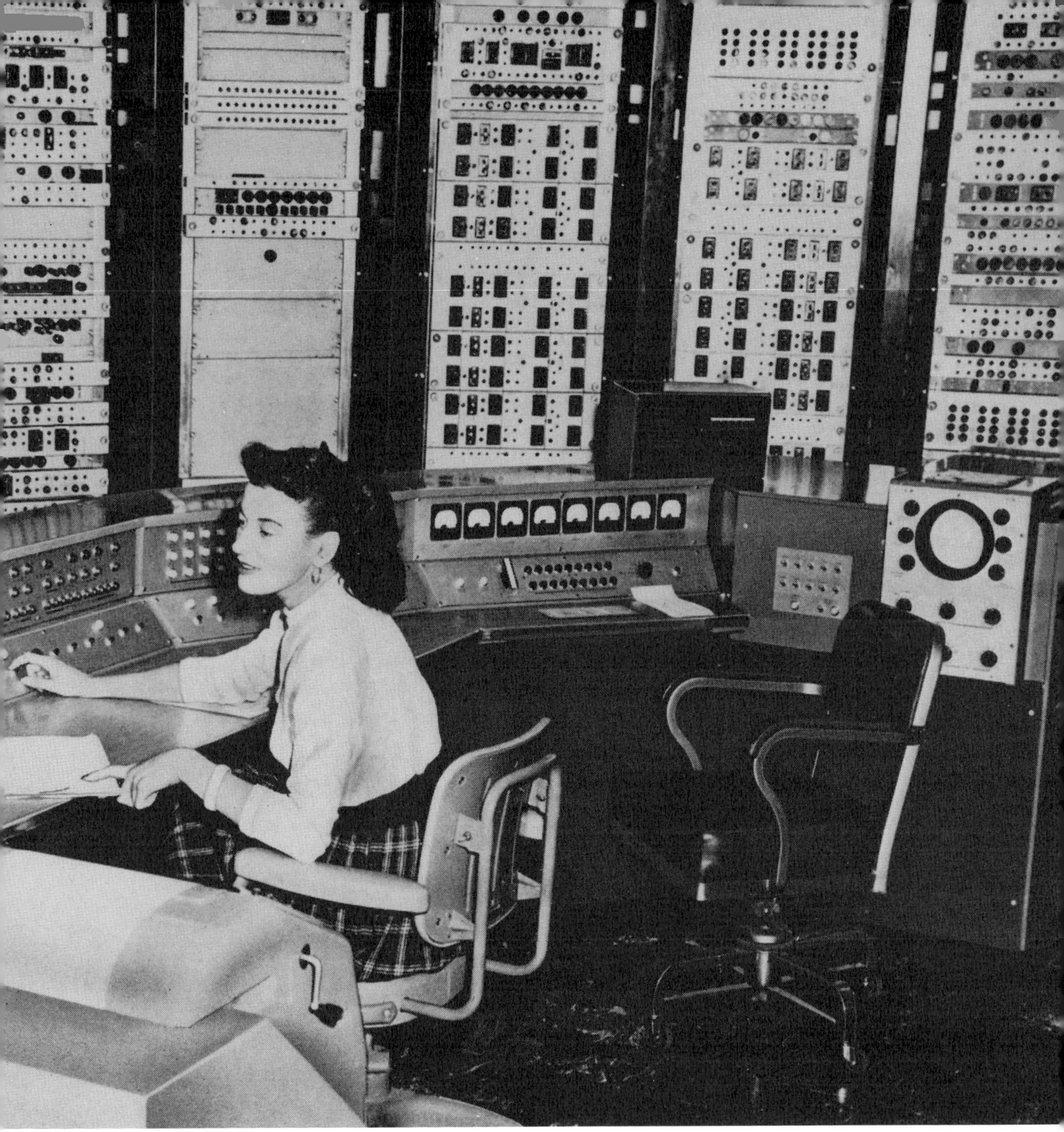

4.5

Raytheon built the RAYDAC for Project Hurricane and installed it at the U.S. Navy's test station at Point Mugu, California, in 1953. (Photo: W. de Beauclair, Raytheon Corporation)

Chapter 5
Whirlwind

All of the projects of the Navy's Special Devices Division depended on computers, but what set them apart was the projects' attempts to use computers in a special way not possible with simpler forms of calculating machinery. They attempted to use the computer not just to augment human powers of analysis but to emulate a person's ability to synthesize or to build models of the world. These projects each attempted in one form or another to simulate the physical world of flight.

Simulation

Simulation describes an activity that pervades aerospace, and so it defies easy definition. Simulation begins by constructing a model of a physical process, but that is only the first step. What gives simulation its power is that this model is dynamic: it simulates the physical process in space over a span of time. Consider the example of an architect's scale model of a proposed new building. From this model one can learn a lot about how the finished building will look, how it will fit in with its neighbors, and so forth. But that model does not say much about how the building will behave under different weather conditions, or how long a lifetime the building will have. The need to add the time dimension is acute in aerospace, which works with a dynamic and fast-changing environment.

One can build scale models and give them a dynamic quality by modeling the dimension of time as well. (The U.S. Army Corps of Engineers has built such a model of the Mississippi Basin, which they use to simulate floods, dams, and the like. Wind tunnels, described in chapter 7, are also such models.) One can also build a model that is a set of mathematical equations describing physical or chemical activities known to occur. But equations alone, without the dimension of time, describe only part of the phenomenon. An equation of an ellipse is a good model of the orbit of Mars around the sun, but it is not a simulation.

Here is where the computer comes in. A computer has the ability to bring the mathematics to life by solving those equations again and again over a span of time. (Indeed, a good definition of a computer is a machine that constructs dynamic models.) When the equation of an ellipse is incorporated into a computer program with information about other parts of the Solar System, it becomes a red disk moving across a computer screen in a path and at a speed proportionate to those of the planet Mars and so is a good simulation.

For the projects of the Navy's Special Devices Division, simulation offered a way to manage the complexity of the postwar air-flight environment. It enabled the Navy to explore and observe the behavior of new types of craft when traditional methods of analysis were of little help. Project Cyclone simulated the characteristics of a missile in flight. It employed an analog computer with a set of special circuits that produced signals according to the solution of up to ninety equations of motion needed to describe the missile. Project Typhoon simulated the interception of an enemy bomber by a missile. It used a similar analog computer to drive tiny scale models of the bomber and missile in a cage the size of a telephone booth. By experimenting with the values plugged into the various equations, one could see whether the two would collide.[1]

The Whirlwind Computer

Project Whirlwind had a different goal. That was to simulate the flight characteristics of an airplane before it was built and to use that information to train flight crews to be ready to fly the plane when it entered service.

Whirlwind had its origins in Luis DeFlorez's wartime plan to build a general-purpose simulator that the Navy could easily reconfigure for several different aircraft then being designed. Such a device would enable the Navy to get a head start on learning how airplanes behaved and on training crews for them without building a special trainer for each one. In December 1944 the Navy contracted with MIT's Servomechanisms Laboratory to develop an Airplane Stability and Control Analyzer, which would embody DeFlorez's concept.[2] In its initial design, the simulator would use a computer made up of a set of standard analog circuits that solved the basic equations of flight dynamics. A plugboard similar to a telephone switch-

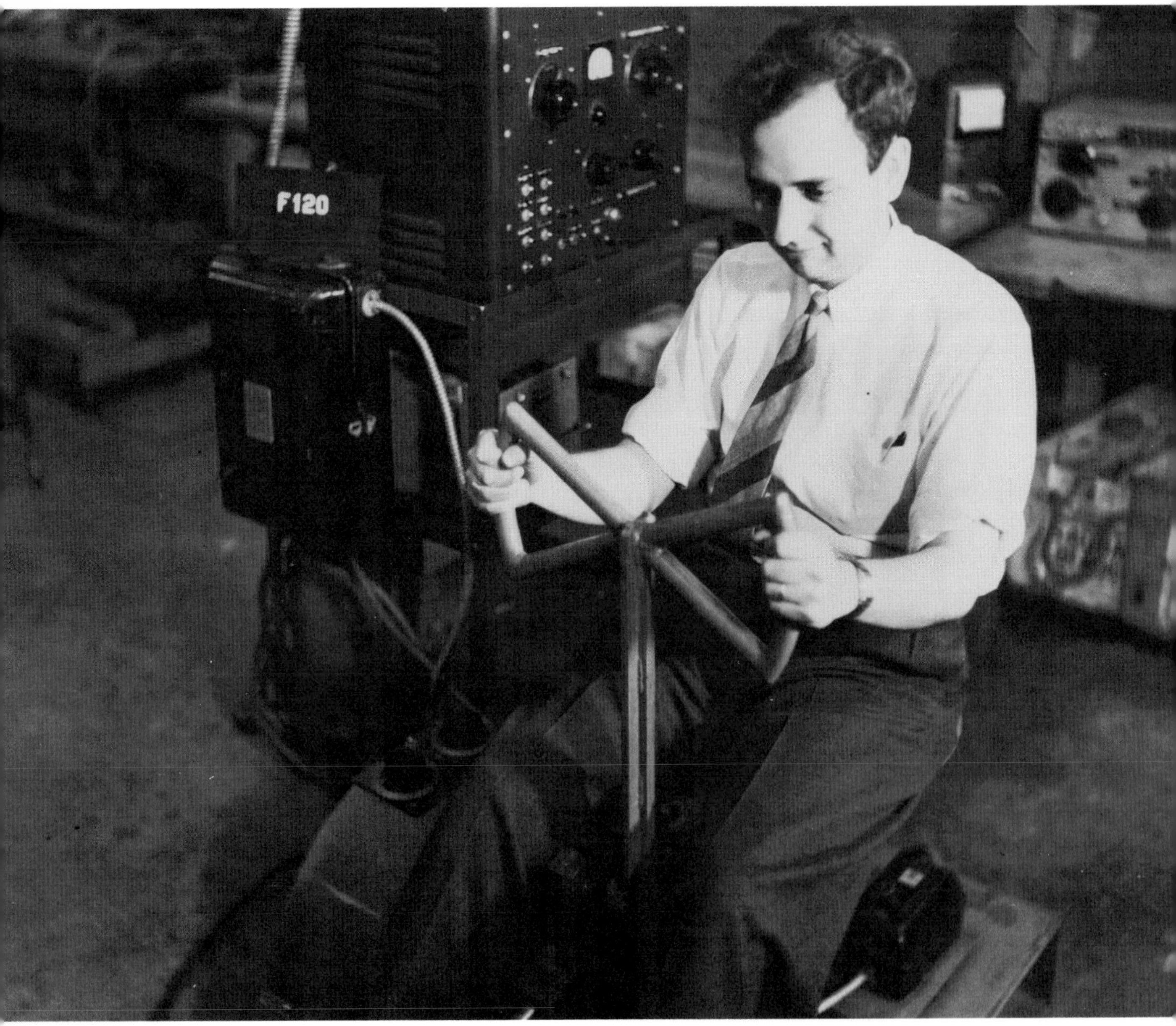
F120

5.1

board would allow aircraft designers to reconfigure these circuits to model the behavior of each new aircraft. The computer would produce results in several forms: as tables of numbers, graphs plotted on paper, and especially as signals to a mock cockpit that recreated the instrument reading and even the feel of the controls of the simulated airplane (figure 5.1).

By the summer of 1945 an Allied victory was assured, and the pressure to train pilots eased. But the Navy continued the project with the Massachusetts Institute of Technology. In March 1946 it became one of the Special Devices Division projects and was given the name Whirlwind. By the beginning of that year the project's director, J. Forrester, had already made the decision that the computer should be a general-purpose digital machine based on the stored-program principles developed at the Moore School by Eckert, Mauchly, and von Neumann. Such a machine could be used to solve a variety of problems besides those of training pilots, a task less urgent after 1945. With a fast enough calculating speed the digital computer could, of course, serve as a simulator. A simulator, the Whirlwind team noted, solves the equations of motion to *substitute* for the effects actual forces of air and gravity have on an airplane; the computer they were building could solve those same equations to *predict* and *analyze* those same forces.[3] That was Forrester's insight. He recognized that the Navy could employ a well-designed, general-purpose digital computer for almost any need it might have.

Whirlwind came close to being canceled, but in March 1950, a few months after the Soviets exploded their first atomic bomb, the U.S. Air Force took over as a major supporter. The Air Force would use Whirlwind not as a flight simulator but as the nerve center for an air defense system for the United States (figure 5.2).[4]

The transformation of Whirlwind from a simulator to a stored-program digital computer had its price: the original estimated costs of $850,000 for the entire project ballooned into annual outlays of about $1 million a year from 1947 through 1953

5.1

Robert Everett sitting at the controls of the Aircraft Stability and Control Analyzer in June 1947. (Photo: MITRE Corporation Archives)

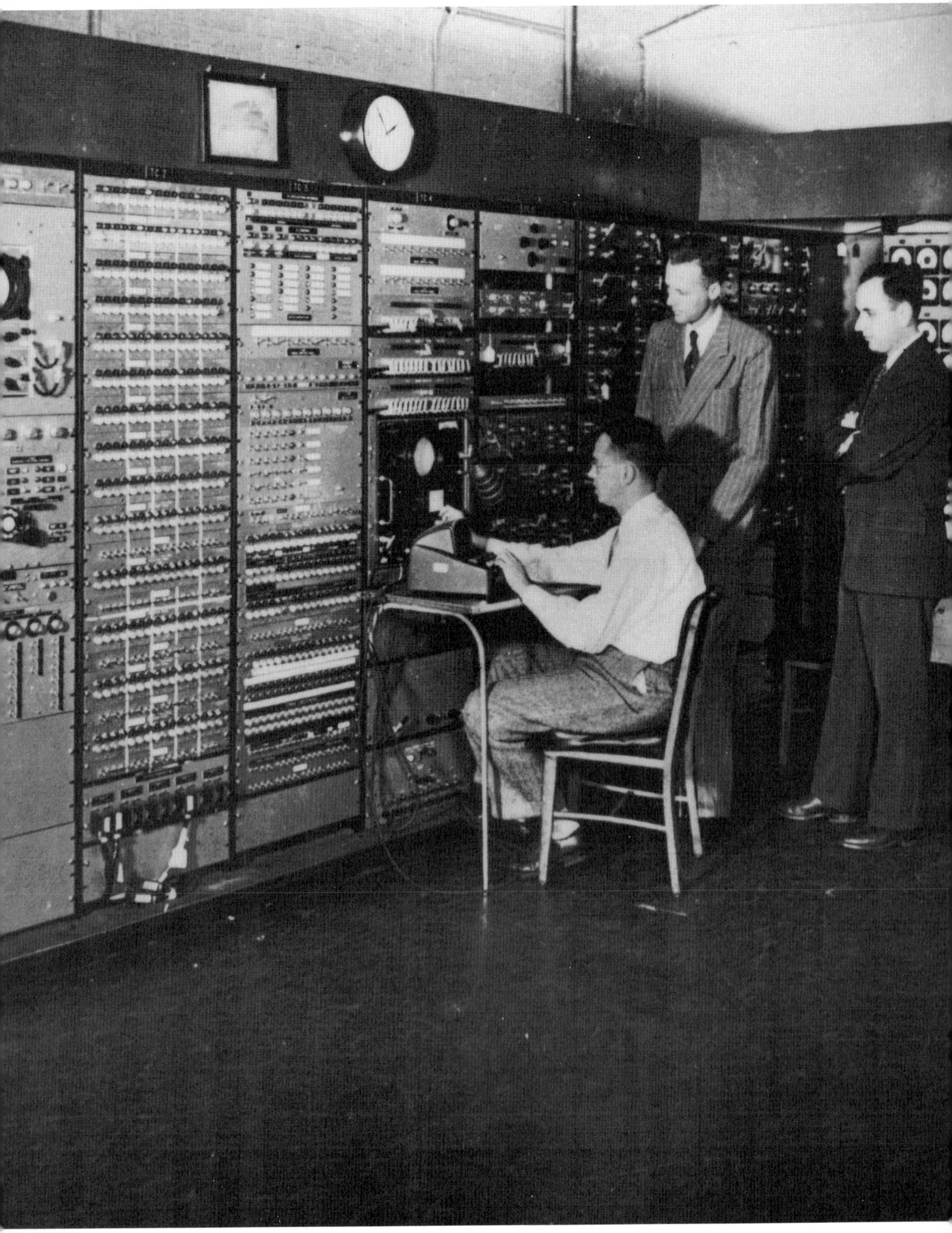

5.2

The Whirlwind console room, ca. 1950. J. Forrester and Robert Everett are standing before the
main control panel of the computer. (Photo:
MITRE Corporation Archives)

(compare this with the cost of $200,000 for SEAC or $500,000 for the Hurricane computer). Much of this cost came from Forrester's insistence on high reliability, something that few of his contemporaries understood. Another reason was the need to have the Whirlwind computer operate at high speeds.

The Air Force wanted a system to take data via telephone lines from a number of radar stations and compare those data with internally stored information about which airplanes were known to be in those locations. The computer would then determine the identity, location, and direction of all aircraft found, and display that information to Air Force personnel in a form they could comprehend. The system had to operate fast enough to allow for a response if enemy planes appeared on the screens. Experience with the Project Hurricane computer showed that a computer needed a storage device that could store and deliver numbers as fast as its processor operated on them. Whirlwind had to be fast all around; the success of the whole system depended on it.

The Whirlwind computer took form in a building near the MIT campus in Cambridge, and in 1951 it reached a stage where it could solve test problems. For its memory device Forrester employed large vacuum tubes, which stored binary numbers as electrostatic charges. This device gave the computer a performance as good as any other computer in use at the time, but it was not fast enough. Even as the computer began operating, Forrester developed the new technique of storing numbers on tiny cores of magnetic material (figure 5.3). After an heroic engineering effort (and more expense), a core memory was attached to Whirlwind in 1953.[5] The Whirlwind team also developed a large television tube that displayed radar information and allowed an operator to query the computer about what was on the screen. These devices combined with a high-speed processor to make Whirlwind fast enough for real-time operations.

With core memory and an input-output screen the electronic digital computer came of age. At least it had the speed, reliability, and computational power that its inventors had been

5.3
Whirlwind engineers began experimenting with
magnetic-core memories in 1950. By 1953 work
had progressed enough so that core planes could
be wired into the computer to replace the storage
tubes. J. Forrester is here examining a 64 by 64
bit memory plane. (Photo: MITRE Corporation
Archives)

5.4

The SAGE computer was the nerve center for a complex system of radar installations, aircraft, ships, and command centers. SAGE was the beginning of all systems hardware and software.

Airborne early warning plane
Picket ship
Higher headquarters
Radio
Interceptors
Texas tower
Radio
Nike
Search radar
Antiaircraft installations
Direction center
Gap-filler radar
Other missiles
Height-finder radar
Weapons status
Adjacent direction center
Ground observer corps
Weapons bases
Flight plans
Weather data
Civil defense

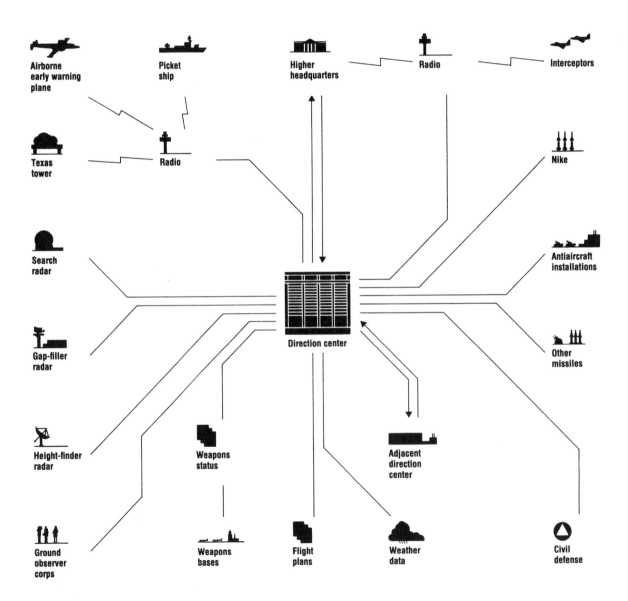

promising for almost a decade. Companies like NCR, RCA, Burroughs, and others that had earlier been reluctant to offer large and expensive commercial computers now were quick to introduce models that incorporated the engineering features of Whirlwind, especially its core memory.[6]

SAGE

The Air Force got the performance it wanted. In 1952 it conducted tests of an air defense system using Whirlwind linked by telephone to several radar stations on Cape Cod. The results were encouraging. With the faster memory access that core memory provided, the Air Force initiated a project to develop a production version of the computer for air defense. The IBM Corporation was the chief contractor, and the computer (called AN/FSQ-7, for "Army-Navy Fixed Special Equipment") became the heart of the Semi-Automatic Ground Environment (SAGE) system of air defense, which monitored North American skies from the late 1950s until 1982 (figure 5.4).

SAGE became fully operational in 1963, with pairs of Q-7 computers installed in concrete buildings at twenty-six sites in the United States and Canada (figure 5.5). The computers themselves were easily the largest single-processor computers ever built. Each machine had upward of twenty-seven thousand vacuum tubes, and each pair consumed three million watts of power. The total cost of SAGE has been estimated at $8 billion, a significant fraction of the U.S. defense budget in those years.[7] A debate continues over how much the expensive SAGE system contributed to U.S. security during those years of cold war (figure 5.6). The costs seemed justified at a time when people still remembered Pearl Harbor and the failure to heed radar warnings of the Japanese attack.

SAGE never had to help intercept any Soviet bombers. It *simulated* the interception of aircraft and performed other similar operations, and for that it was an enormous success. Although SAGE was not the first flight simulator for training pilots, its simulation of these operations and training of ground crews vindicated DeFlorez's initial idea.

a

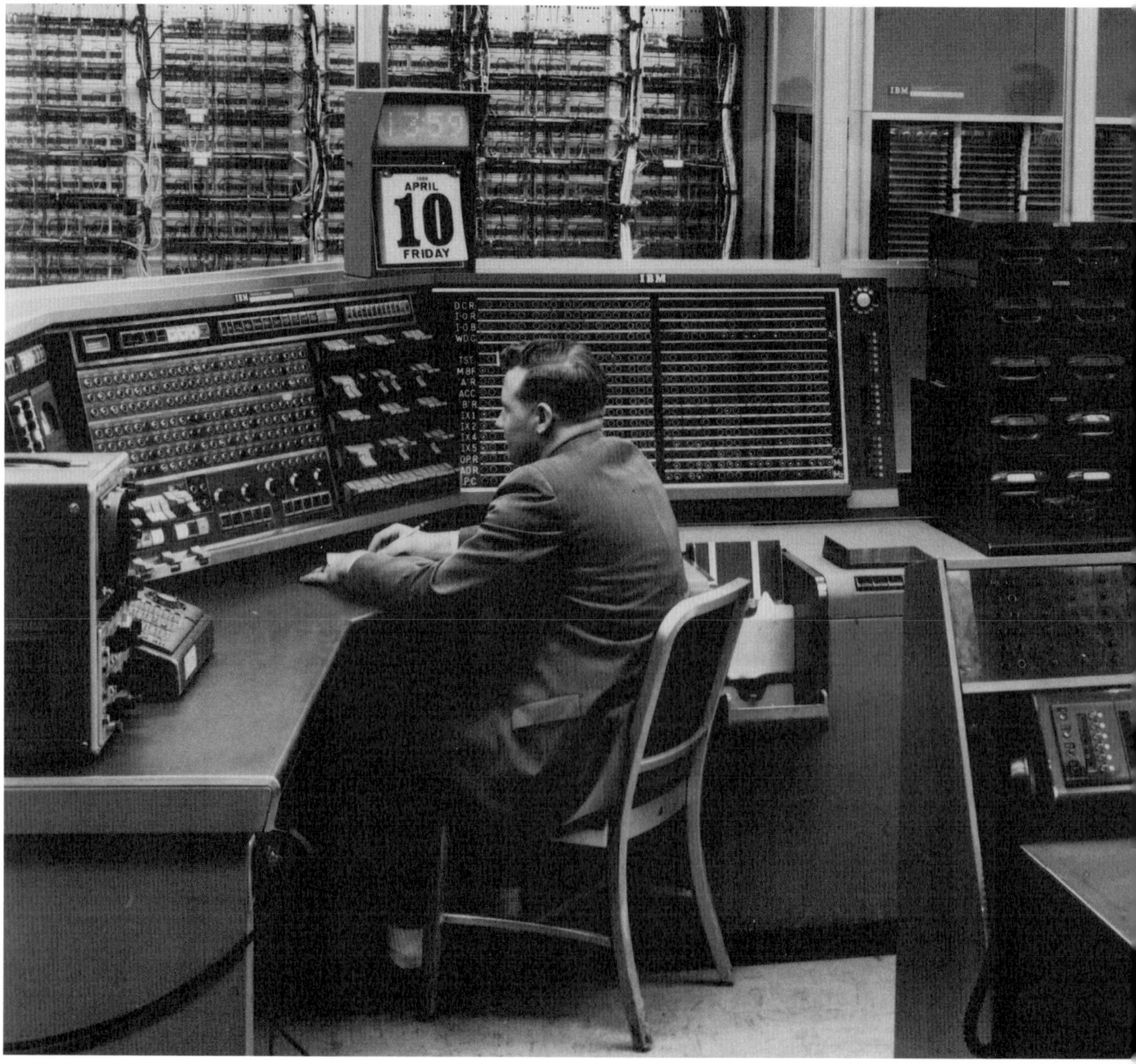
APRIL
10
FRIDAY
IBM

b

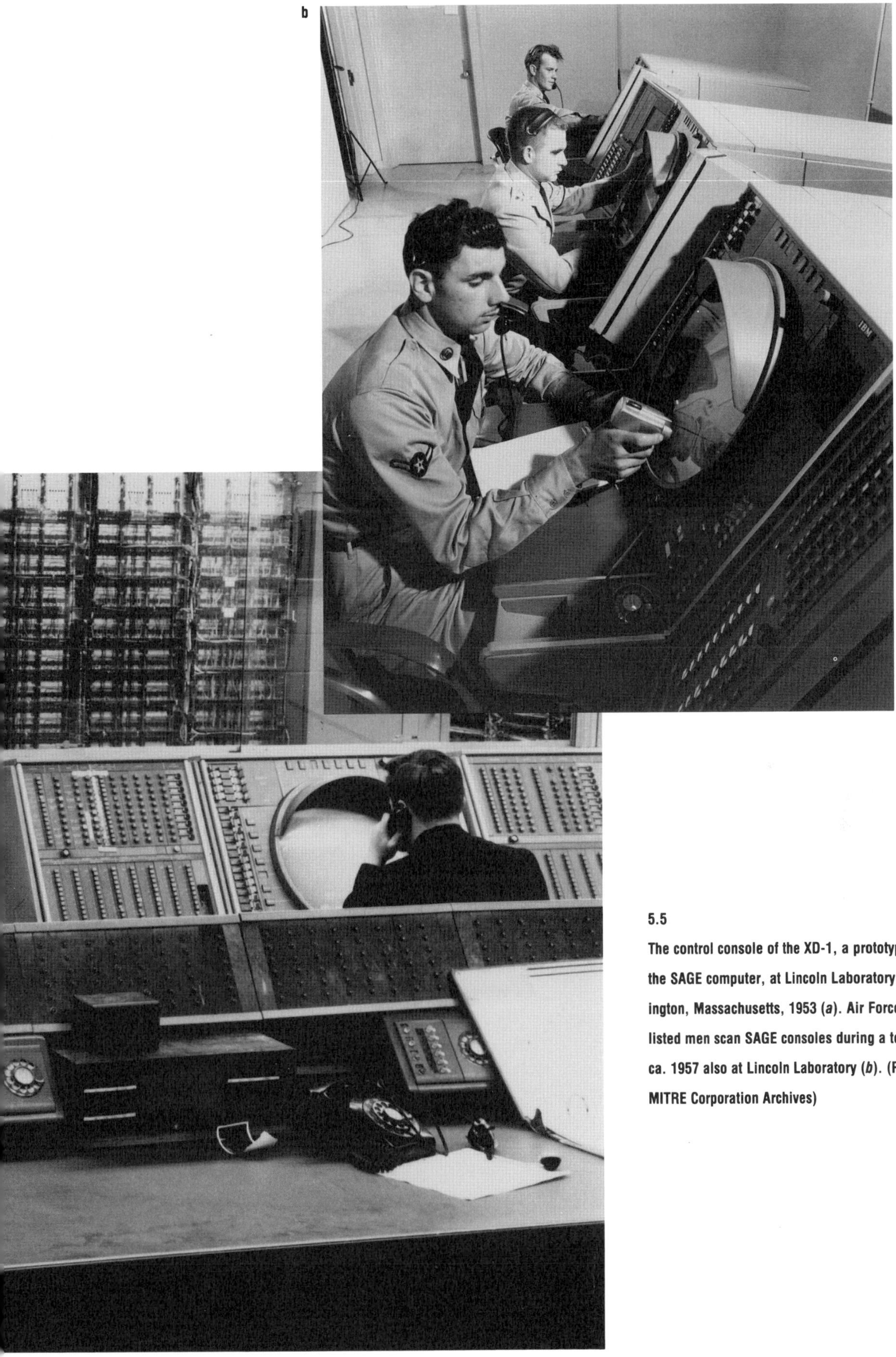
IBM

5.5

The control console of the XD-1, a prototype for the SAGE computer, at Lincoln Laboratory in Lexington, Massachusetts, 1953 (*a*). Air Force enlisted men scan SAGE consoles during a test run ca. 1957 also at Lincoln Laboratory (*b*). (Photos: MITRE Corporation Archives)

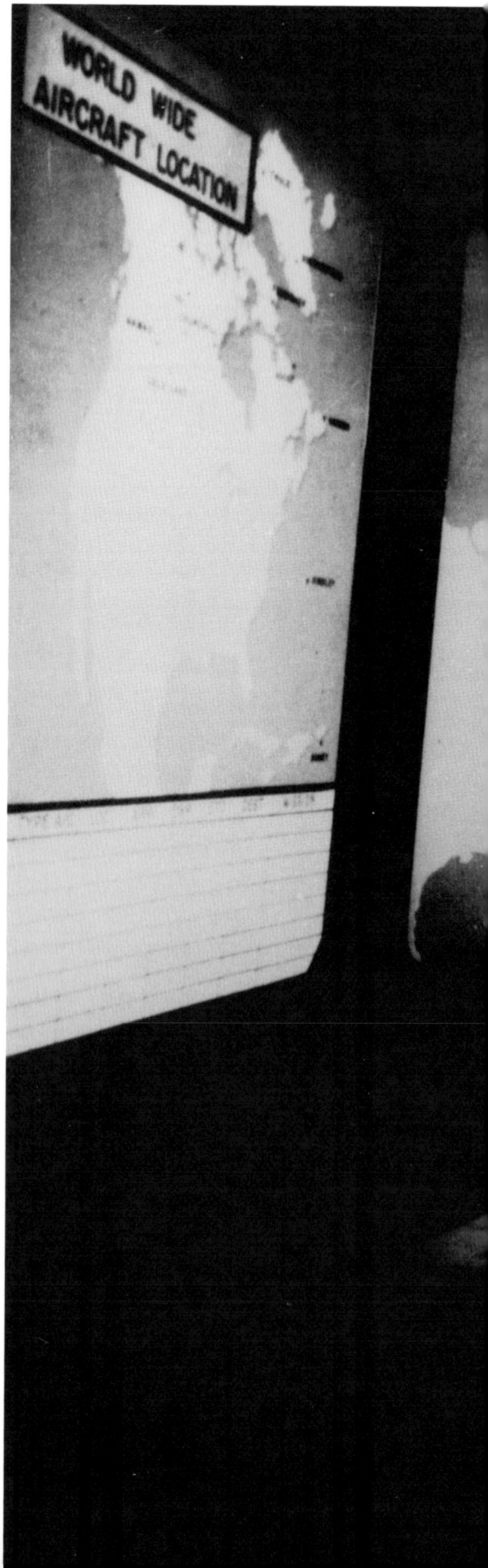
WORLD WIDE
AIRCRAFT LOCATION

5.6

SAGE inspired much current activity in the military for the study of C^3I (command, control, communications, and intelligence). This activity is here exemplified by this 1980 view of Strategic Air Command Headquarters in Omaha, Nebraska. (Photo: Wide World)

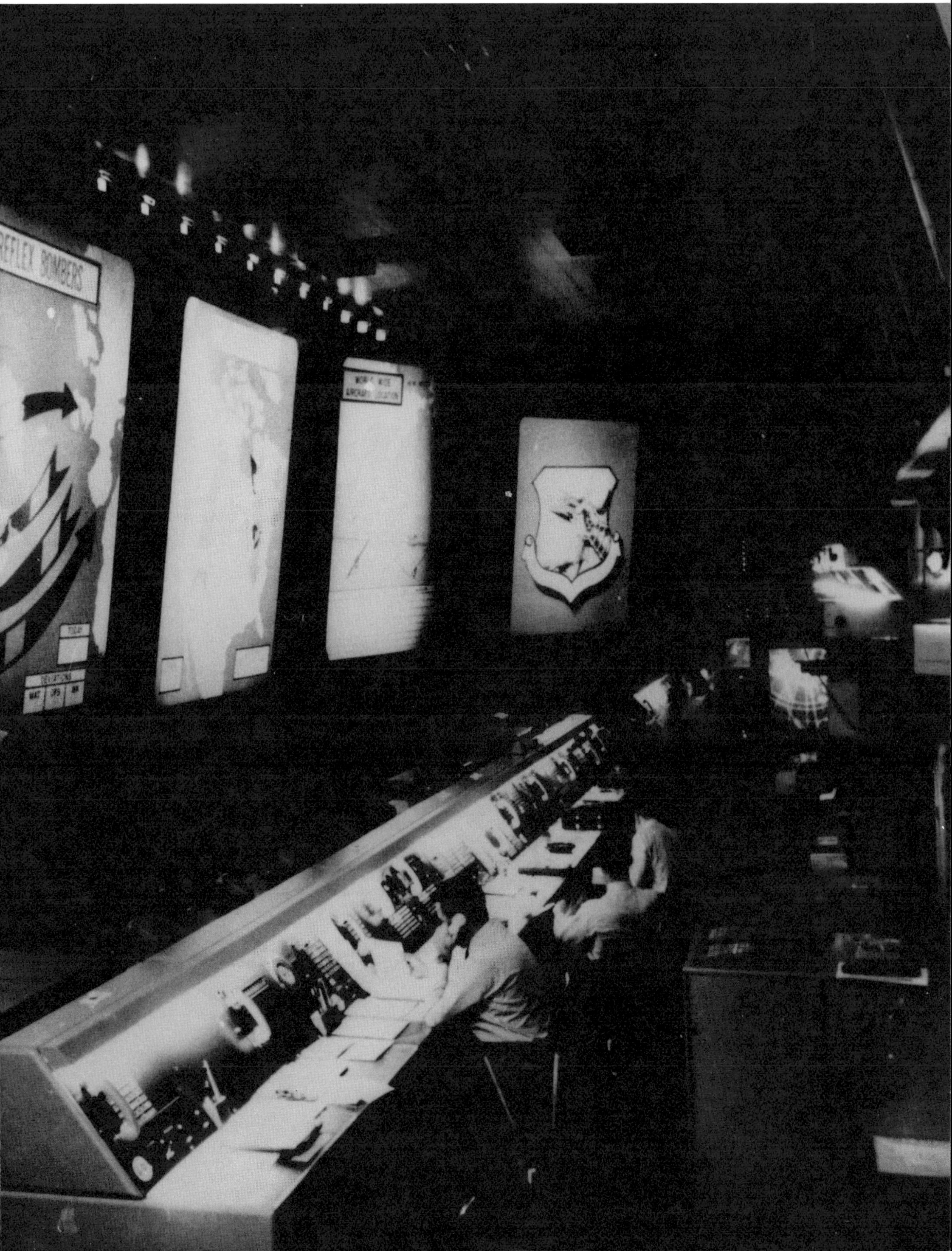
REFLEX BOMBERS

SAGE led to numerous other aerospace applications, of which two deserve special mention here. Experience with SAGE led to the modern system of civilian air-traffic control. This system had the same basic structure as SAGE: radar installations, telephone lines, and computer screens. Indeed, the problem of air-traffic control, to keep aircraft from colliding in the sky, is the mirror image of the problem of detecting and intercepting enemy aircraft.

A second spin-off was for airplane reservations. The high-speed, random-access memory that SAGE required to store and retrieve information about aircraft was well suited to storing and retrieving information about available seats on passenger planes. The airlines and railroads had been experimenting with semiautomatic systems, but not until the IBM Corporation worked with American Airlines to develop a descendant of SAGE called SABRE did a company have an on-line system that gave the ticket agent direct and rapid access to information as needed. SABRE went through an initial period of expensive development, but once established, it was a great boost to American Airlines' business and was eventually copied by most of the other major carriers. The modern, deregulated commercial air-line business would be impossible without SABRE and its offspring.[8]

There is no question that Whirlwind and its production model A/N-FSQ-7 were beneficial to the computer industry as well. The Air Force's demands for reliability, rapid storage and retrieval, and rapid input and output of data led directly to commercial computers with these abilities. For the IBM Corporation, the need to produce and test huge numbers of memory cores gave them an advantage over their competitors that helped them maintain their leading position in the computer industry from the 1960s to the present day. Writing the programs for the complex jobs SAGE needed its computers to perform laid the groundwork for the profession of Computer Programming. (The RAND Corporation spun off a subsidiary, System Development Corporation, for writing the programs for SAGE.)

Finally, at the Massachusetts Institute of Technology experimental programs using Whirlwind led to innovations in simulation, computer graphics, and computer-assisted manufacturing, innovations that would transform the aerospace industry in the 1980s. These innovations will be discussed in later chapters.

By 1960 the forces of the commercial marketplace became the main impetus for the computer industry. Aerospace manufacturers and the air divisions of the Defense Department continued to fund the development of specialized computers for on-board control of air and space craft, but for most engineering applications the marketplace offered computers with sufficient power to satisfy their needs.

With SAGE, aerospace finally had a system that was fast, reliable, and conversant in a graphic language that an airman could understand. And with SAGE the computer industry (especially IBM) learned how to produce high-performance machines on a serial basis in decent production runs. Thus ended the phase in which the computer industry and aerospace grew in lockstep with each other. The computer industry was robust and healthy. It owed its health to customers who, unlike those in Babbage's day, struggled with the complexity of this invention. Federal agencies, including the Census Bureau, the Army, and the intelligence agencies were among the first customers to lend this support. But the customers from aerospace were at least as active and influential, if not more.

Chapter 6

Minuteman, Apollo, and the Chip

For all that SAGE was to aerospace, there was one thing it was not: small. As welcome as the digital computer was to ground-based engineering and operations, the need for its abilities on-board was greater. Digital computers were too big and too heavy. Northrop's experience with the Snark's guidance system made that clear. When finally operational in 1959, the Snark used an analog guidance system, as MAD-DIDA (never mind BINAC) was too big. A computer like AN/FSQ-7, which SAGE used, satisfied none of the on-board needs of airplanes and rockets. Building computers out of electronic components gave them high speed and flexibility. But with electronics came the problems of size, weight, and fragility associated with the vacuum tubes and hand-soldered interconnections of electronic circuits.

Why Small?

The alleged revolution that computers have wrought for aerospace and other parts of American society is due as much to the miniaturization of its electronic circuits as to any other factor. But how and when did computer and electronics engineers recognize this fact? The benefits seem obvious, but only with hindsight. It is not, after all, a universal maxim of engineering that small, dispersed systems are better than large, centralized ones. Consider the philosophy behind the engineering of hydroelectric dams, tanker ships, and urban development projects, for example.[1]

Nor is microelectronics a triumph of the philosophy that small is beautiful. If anything it represents the opposite. In contrast to electronics engineering in the vacuum tube era and contrary to the myths that have grown up around the origins of the personal computer industry, very little in solid state electronics can be done in one's basement or garage. The very design of such circuits requires large computers; fabricating them requires expensive and exotic "clean rooms," far more dust-free than the best hospital room. In these rooms exotic materials are handled at precise temperatures and pressures. Though the circuits themselves are small, the infrastructure needed to make them is anything but small.

The first experimental digital computers of the 1940s were large systems consciously planned that way. And for at least a decade thereafter one school of opinion argued that the best way to exploit the computer was to build a very large, centralized system and then feed its computing power via cables to consumers at home or in the office. (Aircraft can receive computer power sent by radio. But this idea never caught on, mainly because of the difficulties of ensuring continuous radio connections. Something like it was employed for the American and Soviet manned space missions of the 1960s and 1970s.)

But centralized computing was not the wave of the future. The invention of the integrated circuit made it possible to build computers that were small, lightweight and cheap enough that computer power became embedded in other systems, giving those systems an intelligence they hitherto lacked. The guided missile was but one such system that benefited from this innovation.

The historical record shows a plain correlation between the need for lightweight guidance and the invention of solid-state, integrated electronic circuits. But aerospace was not the sole driver of this invention. Bell Laboratories' inventors developed the transistor with neither aerospace nor even the computer in mind; they simply knew that there was a market for a solid-state electronic circuit that could do the work of vacuum tubes. Electronics engineers and their consumers, both civilian and military, argued for miniaturization and lightweight systems. In contrast to, say, electric power engineering, those in communications and computer technology who argued for miniaturization were a strong, articulate group.[2]

As inventors, entrepreneurs, and financiers began building electronic networks for communications in the 1920s and 1930s and for computing systems in the 1940s and 1950s, they realized that almost every other technology or system could benefit from the intelligence that electronics offered. That is, electronics began to be perceived not so much as a separate industry but as the brains and nervous systems of other machines, whatever they might be. Every machine needs a source of power, a mechanism for translating that power into useful forms, and a control mechanism, a way to get that machine to do what we want it to do. These requirements hold for machines both large and small. Human and animal muscles supplied the power for machines for millennia; up until the twentieth century humans (with a few exceptions) had to provide all the control.

Thus, the smaller and lighter one could build computers, the more they could act as controllers for other technologies. For aircraft, with their stringent weight and space requirements, the need to miniaturize was even more compelling. And for unmanned ballistic missiles and spacecraft, miniaturized, on-board control was an absolute requirement.

Simon Ramo, a founder of the aerospace corporation Ramo-Wooldridge (now TRW), was one of the more vocal spokesmen for microelectronics in the postwar environment. His company itself is evidence of the importance of computers and electronics for aerospace. Companies like TRW, SDC, and MITRE do not directly manufacture spacecraft but coordinate and manage the various subcontractors and government agencies that do. They perform the systems integration needed for projects of such complexity. The very discipline of systems engineering had its origins in electric-power engineering and later computer engineering, and companies that perform it for aerospace are staffed primarily by people with training and experience in electrical engineering and computer science.

In 1962 Ramo summarized the relationship between missiles and microelectronics as follows:

While in most guided missile and space vehicle systems the electronics—the "brain" and "nervous system" of the whole beast—constitutes a very small fraction of the total size and weight, the burden for miniaturization nevertheless falls heavily on the electronics engineer. There are several reasons for this.

One is that, in general, the electronics equipment of all airborne or spaceborne systems must go to the "end" while, in contrast, a good deal of the initial weight, notably fuel and oftentimes major portions of the propulsion system, will have been exhausted and purposely dropped along the way to lighten the load. The guidance and control must be present to the last moment where any unnecessary weight exacts a severe penalty.

Another reason for the emphasis on miniaturization of electronic equipment is that it can *be miniaturized, and any weight that can be saved must be saved.*"[3]

As mentioned in chapter 2, aerospace engineers will shave weight from anything they can.[4] They make every structural piece as thin as possible while maintaining a respectable safety factor. Fuel weight, which in a rocket constitutes by far the greatest portion of total weight, can be reduced by using liquid hydrogen, a very light fuel but tricky to work with. Manned missions have absolute limits imposed by the human body and its need for space, air, water, food, and so forth. (All other things equal, a short, light-framed person has a better chance of becoming an astronaut than a hefty six-footer.) That leaves guidance, control, and communication equipment as the place where weight and size reductions must take place.

The postwar trend toward ever more complex craft with more ambitious missions and higher speeds meant that these on-board systems had to do much more. By 1957 electronic components accounted for up to half the cost of a missile, and a significant portion of employees at most aircraft companies were electrical or computer engineers. This trend accelerated rapidly after Sputnik (figure 6.1).[5]

The close relationship that characterized postwar computing and aerospace repeated itself in the 1960s in the area of missile guidance and control. As Southern California aircraft companies drove the computer industry in the 1950s, so Air Force and NASA guidance needs drove the development of digital circuits in the 1960s. In both cases the effect was to move experimental, fragile devices out of the laboratory and into mass production. Guidance needs went further: they drove down the costs of circuits to a point where they became cheap enough to go into consumer products. Guidance computers gave modern society the ten-dollar calculator, the Walkman, the home videotape player, and the home computer.

6.1

The growing complexity of avionic equipment aboard bomber aircraft. (Based on statistics from USAF Investigates Basic Molectronics, *Aviation Week,* 71 [August 17, 1959], p. 77)

130
120
110
100
90
80
70
60
50
40
30
20
10
Number of electronic components
(in thousands)
B-17
B-29
B-50
B-47
B-52
B-58
1940
1945
1950
1955
1960

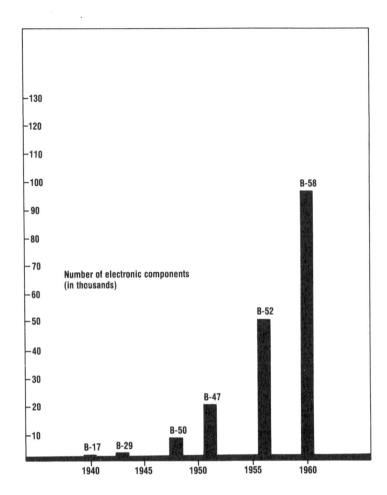

Inertial Guidance

Missile designers were especially fond of the inertial method as the best way to guide a rocket through space. Though in practice a complicated thing to accomplish, inertial guidance is based on a simple, direct application of Newton's laws of motion, which state the relationship between a force acting on an object and any changes in the speed or direction of that object.

It is impossible to discern from inside a vehicle moving at a constant speed (e.g., an elevator) how fast it is going or where it is. To do that requires reference to a point outside the vehicle. But *changes* in motion, acceleration, can be felt and measured (as when the elevator slows down as it comes to a floor). A pendulum or spring will record a change in speed or direction and thus can serve as the basis for an instrument that measures the acceleration of an object. By summing up these changes over an interval of time, one can compute the speed and direction of an object at any moment. And by summing up these quantities once more, one can know the position of an object.

Inertial guidance is sometimes called astronomy in a closet, as it seems to offer a way of navigating without ever referring to anything outside, be it the ground or the stars. For an inertially guided missile to do that, it must know with great accuracy both its initial position at launch and the effects of gravity, which the system must subtract from its acceleration measurements. (A large portion of the Air Force's ballistic missile budget is spent on measuring the shape and gravity of the earth to a high degree of precision.)

In the late 1940s and early 1950s advocates of inertial guidance had to convince skeptical experts that such a thing was, indeed, possible. To get such a system to work required an improvement of at least four orders of magnitude (10,000:1) over existing autopilots and other devices then in use. Some argued that inertial guidance was in principle impossible, since one cannot distinguish between an acceleration due to motion from one due to the earth's gravity.[6] Gravity thus poisons the readings of any inertial system's measurements. By about 1953 it was clear that both objections could be met: the first by a brute-force engineering assault on the components, the second by adopting the ingenious technique of carrying a computational model of the earth's gravity along in the guidance system, so that the system could account for those effects.

A typical inertial-guidance system contains three subunits. The first is a set of pendulums or springs that measure acceleration along the three axes of space. The second is a set of gyroscopes that provides a stable reference point or a stable platform against which to measure accelerations. By taking advantage of the tendency of a spinning gyroscope to resist movement from an initial position, the stable platform provides the basis for calculating accelerations in spite of the pitch, roll, and yaw of the craft. The third is a computer to perform the calculations that yield the position of the vehicle from the data of the accelerometers. The calculations are part of Newton's calculus and are largely a matter of summing the accelerations up twice for the time interval required. The computer must also solve trigonometric equations that convert this information into coordinates of latitude, longitude, and altitude. In an unmanned rocket the output from the computer drives the rocket motors or guide vanes to keep the machine on course. In a manned system the computer can display the information for the pilot.

As in so much of space flight, Wernher von Braun's team at Peenemünde were pioneers in developing the rudiments of inertial guidance (figure 6.2). In September 1939 von Braun addressed a meeting of scientists, engineers, and mathematicians on the problem of guiding the rockets his group had just then begun to develop, and a number of possible solutions were discussed. The basic problem of guiding a ballistic missile was to determine and actuate engine cutoff to an accuracy of one part in a thousand. From the meeting two candidates for a guidance system emerged: inertial guidance and Doppler radar. Inertial guidance eventually won out, despite the technical difficulties it presented.[7]

During the Second World War Walter Haussermann led a team at Peenemünde that developed a guidance system using analog circuits for the V-2. It was a crude device that was nevertheless effective enough to guide the rocket during the first few minutes of flight, which was all that was required, as after the engines cut off, the V-2 coasted to its target. After the war von Braun continued this work in the United States at the Army's Redstone Arsenal in Huntsville, Alabama (now NASA's Marshall Space Flight Center). At that time Northrop was working on the same problem for the Snark, for which they proposed an inertial-guidance system for the final descent of the missile to its target (when celestial guidance would no longer be effective). The two other centers of this activity in the United States were at the Autonetics Division of North American Aviation, also in Southern California, and the MIT Instrumentation Laboratory in Cambridge, Massachusetts, and under the leadership of Charles Stark Draper.

Autonetics and the Instrumentation Laboratory pushed forward with the technology the hardest. These organizations first perfected fully inertial systems and made two pivotal deci-

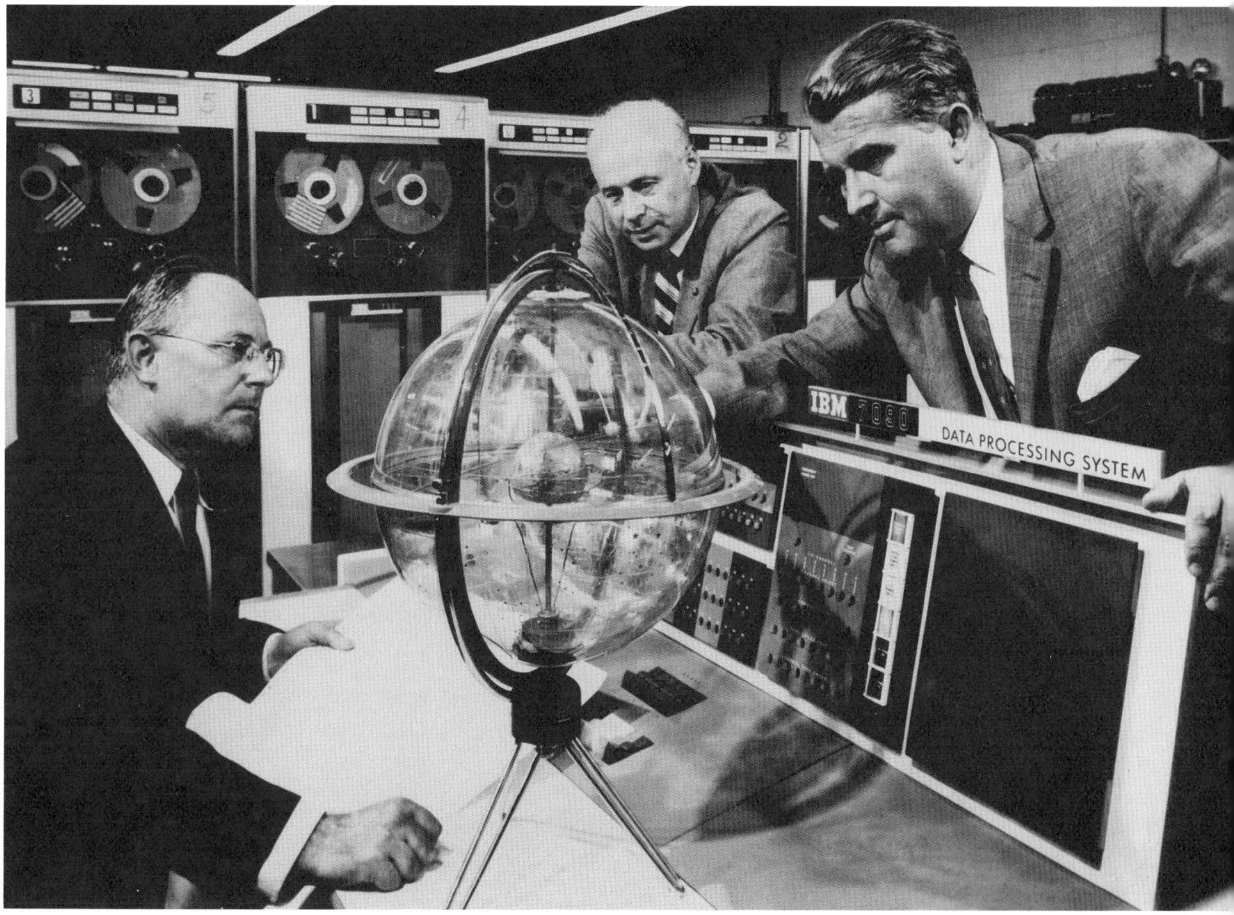
IBM 7090
DATA PROCESSING SYSTEM

6.2
Von Braun's wartime experience with missile guidance served as the basis for his work on the Saturn rockets, which boosted Apollo astronauts to the Moon. In 1960 at the onset of Apollo von Braun and his fellow engineers, many of whom had come from Peenemünde, installed an IBM 7090 computer at NASA's Marshall Space Flight Center in Huntsville, Alabama, to simulate the conditions of launching the massive Saturn rocket to the moon. Here Dr. Wernher von Braun (right), center director, reviews flight trajectories with Dr. Helmut Hoelzer (left), director of the Computation Division, and Dr. Eberhard Rees, deputy director for research and development. (Photo: IBM Corporation)

sions in 1962 that would have far-reaching consequences. Engineers at Autonetics shifted from analog to digital computers for guidance (recall that the Snark used an analog device after an attempt to use a digital computer). And for the U.S. Air Force's Minuteman missile, they adopted integrated circuits, again a first. Across the continent engineers at the Instrumentation Lab reached the same two decisions for guidance and navigation for the manned Apollo Lunar Mission: a digital computer, using integrated circuits.[8]

The Integrated Circuit

An integrated circuit combines a number of individual components, such as transistors and resistors, onto a single "chip" of material, usually silicon. Two men, Jack Kilby of Texas Instruments, and Robert Noyce of Fairchild Semiconductor, invented it within six months of each other in late 1958 and early 1959. Kilby's version came first; it showed that such a device was possible. Noyce's version had interconnections as well as individual components on a slab of silicon and led directly to methods of fabricating complex circuits, especially computer circuits, in large quantities.[9]

Almost any circuit can be etched onto silicon, but computer circuits best suit the technique. The art of making integrated circuits has advanced in recent years to a point where it is now routine to make a chip less than a quarter inch square containing all the circuits needed for a general-purpose computer as powerful as the ones for SAGE, which took up a whole building. Mass production of these chips has driven their price down to around ten dollars for one that can perform general computations. No other technology has advanced so rapidly as computer circuits between 1960 and 1985.

The story of the invention of the integrated circuit is one of the dramatic stories of our time. It has grown to the status of legend. But this legend obscures how the chip, like the digital computer, had a difficult time emerging from the laboratory into the consumer market. The tremendous success of this invention makes its difficulties hard to imagine. After all, chips are obviously such cost-effective, rugged, and powerful devices: was it not obvious that they were the answer to everyone's dreams?

It was not. In 1962 the integrated circuit was but one of many competing techniques for miniaturizing electronic components, and it was not at all the one most favored.[10] The integrated circuit showed great promise, but making it required working at extremely close tolerances, and the yield of good chips was apallingly low. The yield of good transistors was already low: a manufacturing process was considered good if a batch yielded transistors half of which were good. Trying to put more than one transistor on a piece of silicon only made this problem worse.

The result of all this was that although a single integrated circuit could replace up to about a dozen discrete transistors, resistors, and other components, the cost of each integrated circuit was high, around $50 each. Compare this price to $2 to $3 for a vacuum tube and perhaps $20 for a vacuum tube circuit equally complex as an integrated circuit.[11] In this set of circumstances the initial response of the U.S. Navy and Army Signal Corps to Texas Instruments' announcement of their invention was to turn it down. The Air Force too was hostile at first, believing that their own Molecular Electronics program promised an even greater advance over discrete circuitry. But after a while the Air Force realized that its project was making little progress, despite an outlay of millions of dollars mainly to Westinghouse, the prime contractor. They then began placing orders for airborne computer circuits using Kilby's invention.[12]

Minuteman

Minuteman was a three-stage, solid-fueled ballistic missile designed to carry a nuclear warhead from launch sites in the United States to targets in the Soviet Union (figure 6.3). Because its fuel was a solid with fairly stable chemical properties, it could be fueled, placed in a silo, and fired on a moment's notice, hence the name. This gave it an overwhelming advantage over the liquid-fueled Titan, which was stored in its silo fueled with hazardous and volatile liquids. The Air Force let preliminary contracts for the missile in 1958 with Boeing as the prime contractor. Its first test flight was in February 1961, and it became part of the Strategic Air Command's operational inventory in October 1962. By 1965 eight

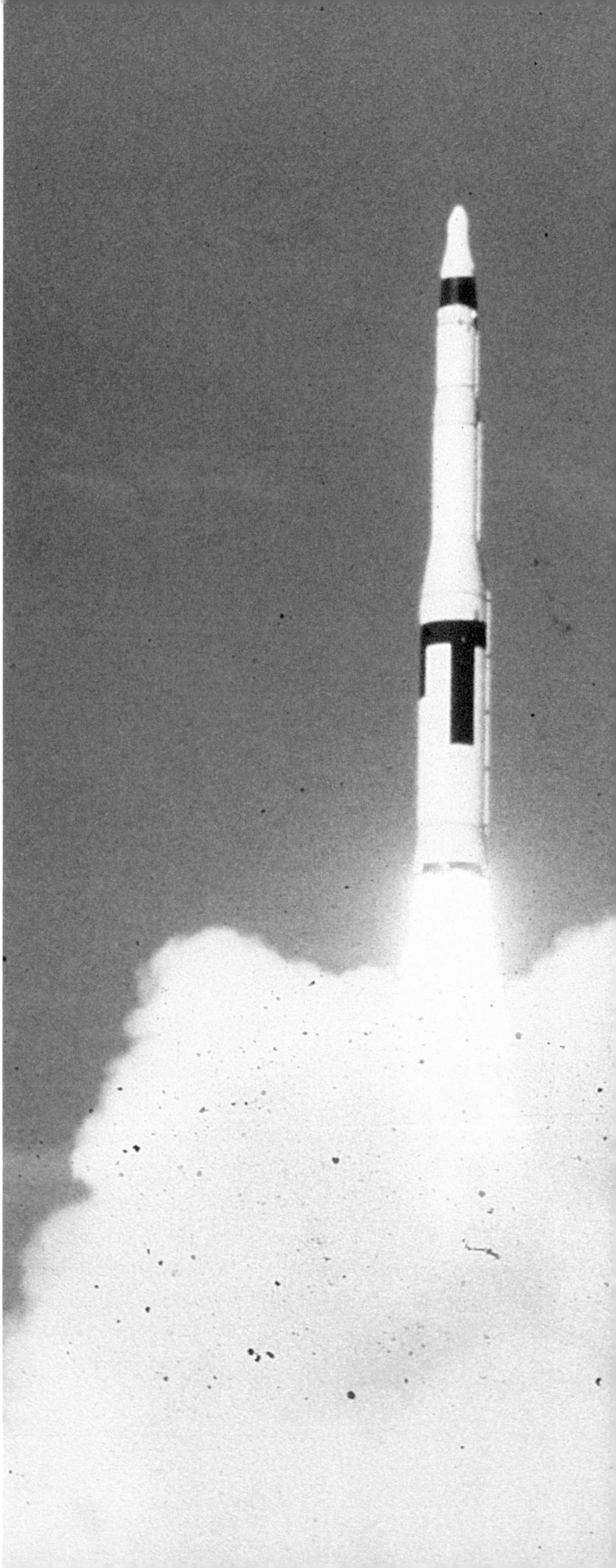

6.3
The test-firing of a Minuteman III (*a*). The Minuteman's guidance system is mounted in a ring between the upper stage and the warhead (*b*). (Photos: U.S. Air Force and David E. Dempster)

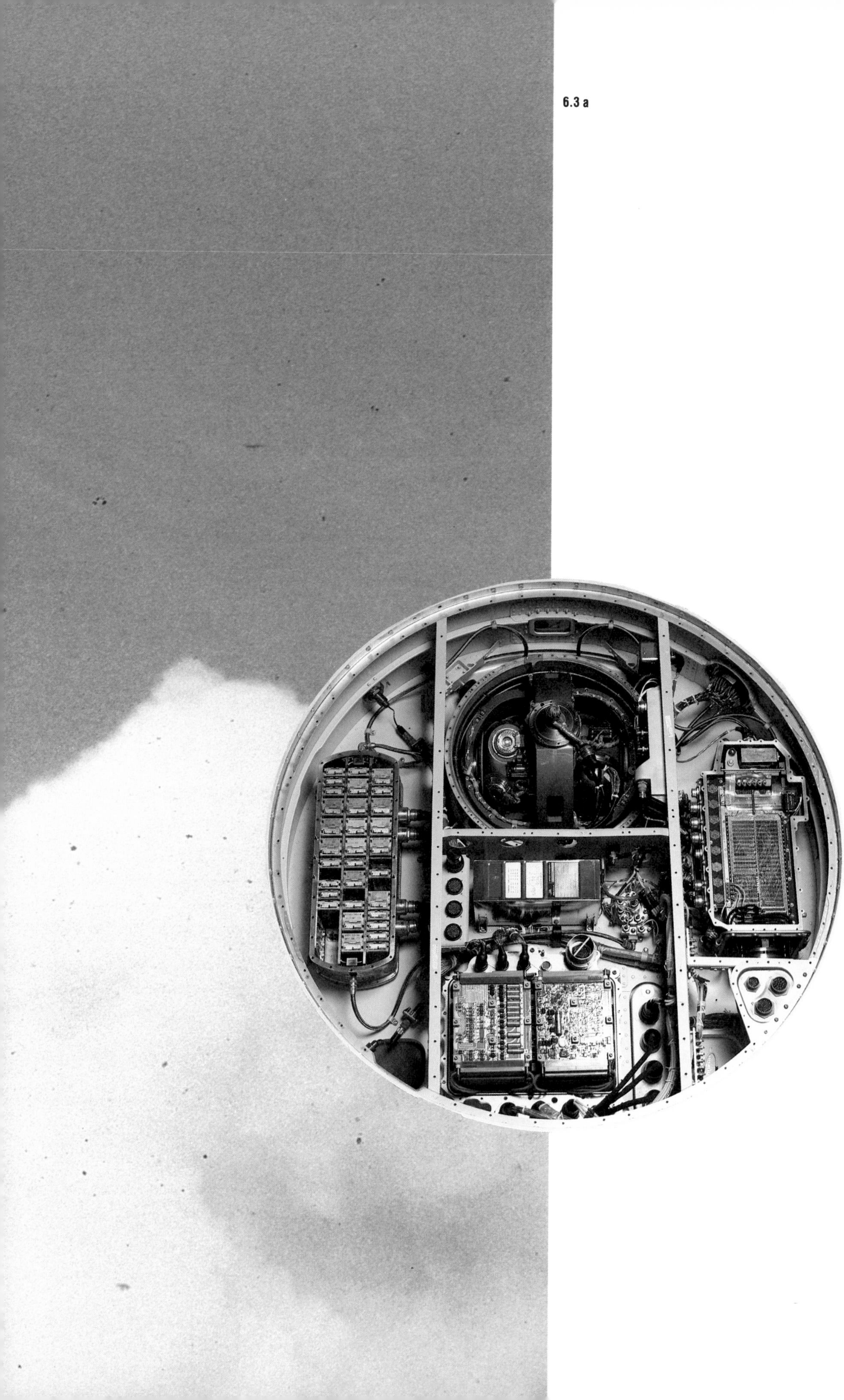

6.3 a

b

hundred missiles were in place in silos at five Air Force bases in the Midwest and Rocky Mountain states. In 1962 work began on an upgraded version called Minuteman II; this was followed by Minuteman III, operational in 1970 and still the mainstay of the U.S. strategic missile arsenal.[13]

The Autonetics Division of North American Aviation developed the guidance system for Minuteman. In 1951 this company proposed a guided missile they called the Navaho, whose characteristics were very similar to Northrop's Snark. The chief differences of the Navaho were its use of ramjet propulsion, which gave it higher speeds, and inertial rather than stellar navigation. The Navaho lost out to the Snark for the Air Force contract, but work on its guidance system gave Autonetics a solid grounding in the complexities of inertial guidance, a basis that made them a leader in the field.[14] For Minuteman a group of sixteen hundred employees under E. N. Ljunggren, C. P. Ballard, and W. L. Morris formed a separate division devoted to the entire guidance, control, and checkout system of Minuteman. In 1962 they chose integrated circuits for the Minuteman II computer.

The V-2 set a pattern among ballistic missiles of using pendulums and gyroscopes coupled to an analog computer for guidance during the brief boost phase. Minuteman I was one of the first to use a fully inertial system that incorporated a digital computer. The guidance system was busiest for the first two or three minutes of flight, yet it had jobs to do for the rest of the flight as well. The guidance computer was also programmed to perform continuous diagnostic checks on the missile as it sat in a silo unattended for years. Such double duty is just the kind of capability a digital computer has and an analog computer lacks. This extra duty fit nicely with the need for Minuteman to be ready to fire on short notice. The Minuteman computer was able to take on this additional job with only a slight increase in missile weight, more than offset by the decrease in cables to an external diagnostic computer that would otherwise have been required.[15]

The Minuteman I computer contained about fifteen thousand components, including discrete transistors. The computer advanced the state of the art with its high reliability for years of unattended operation. The need for such a long time between component failures far surpassed what existing transistor suppliers could deliver, but Autonetics engineers worked with them to produce a computer with an estimated reliable operation of well over six thousand hours before a failure might occur. The standards Autonetics established for accepting components from vendors, for "white room" assembly conditions, and for rigorous testing at all phases of assembly improved avionics reliability by a hundredfold over what it had been.

The Minuteman guidance system worked. But a change in requirements for Minuteman II made demands on it that discrete circuits could not meet. The change came with a change in the U.S. strategic policy from massive retaliation to a flexible response.[16] This meant a need to retarget and thus reprogram the missile up to the instant before launch. Autonetics chose integrated circuits for the guidance computer to meet the computational demands of this requirement.[17]

Texas Instruments of Dallas, where Jack Kilby built the first integrated circuit in 1958, was the principal supplier of circuits (over sixty percent), with Westinghouse supplying most of the remainder. Texas Instruments' first chips sold for $50 to $65 and could replace up to about a dozen discrete components having about the same total value. The main reason for the high cost was the low yield of the manufacturing process: in a good yield half the circuits worked. The reliability of the computer of Minuteman II was poor compared to that of its predecessor. Minuteman I had already set new standards for reliability unmatched by any other electronic system in military or civilian operation at the time. Getting Minuteman II to be reliable within acceptable limits drove the cost of the computer way up.[18]

Autonetics' contract with Texas Instruments eventually did drive the price of integrated circuits down, since it forced Texas Instruments to improve its yield and lower production costs. By 1965 the cost had declined to $12 per circuit. Each Minuteman II computer used about two thousand integrated circuits, and the initial Air Force contract called for a total deployment of hundreds of missiles. (The computer itself weighed thirty-seven pounds, a little over half what the Minuteman I computer weighed.) The Minuteman project nurtured integrated-circuit production through its infancy in the early 1960s; by 1965 the project still was consuming twenty percent of the monetary sales of the industry and was still "the largest single consumer of semiconductor microcircuits," according to the trade press.[19] The decision to use integrated circuits for Minuteman II gave that project a central role in bringing the chip to its present position in society.[20]

Apollo to the Moon

Chips that did not go into Minuteman went into the on-board computer for the Apollo Command and Lunar Modules, which took men to the moon and back. In 1962, the same year that Autonetics chose integrated circuits for the Minuteman II, designers at the MIT Instrumentation Laboratory chose integrated circuits for the upgraded Block II Apollo Guidance Computer, intended for actual lunar missions.

A historic example of fusing computing and flight technologies together is the Command and Lunar Modules of the Apollo spacecraft, which carried men into space and ultimately to the moon between 1968 and 1972.[21] These were the first and only craft to carry men out of the gravitational field of earth to that of another celestial body. The lunar landing, on a body with no atmosphere and one-sixth the gravity of earth, had little precedent in any air or space activities. In Apollo the on-board computers were truly an integral part of the entire system.

In May 1961 President Kennedy challenged the country to put a man on the moon and return him safely "before this decade is out." The administrator of NASA, James Webb, turned to the Instrumentation Laboratory for advice on the critical problem of guidance and navigation. Under the leadership of Charles Stark Draper (1901–1987) the laboratory had established a reputation as a producer of missile guidance systems with their work on the Polaris in the late 1950s. Webb knew he could count on "Doc" Draper to come up with engineering solutions to problems for which there was little or no precedent. And above all, Webb knew that Draper and his laboratory would see an idea all the way through to the timely completion of a working, reliable product, something the Apollo project could not do without (figure 6.4).

6.4
Charles Stark Draper (center) aboard Air Force 1,
with NASA Administrator James Webb (left) and
Lyndon Johnson. (Photo: Charles Stark Draper
Laboratory)

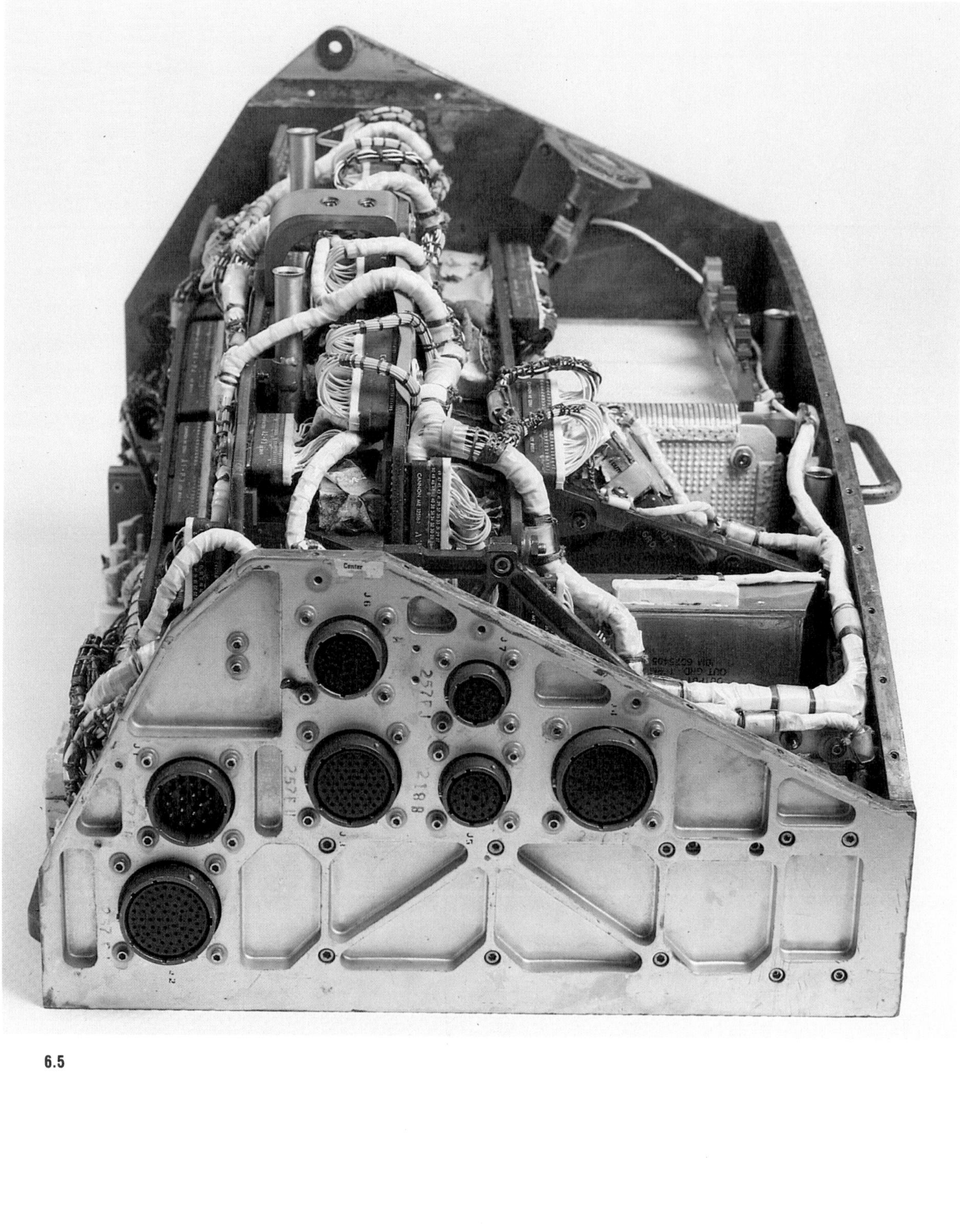
Center

6.5

Draper recounted his meeting with NASA just before the first manned lunar landing in 1969. The question of guidance and navigation for Apollo came up. According to Draper,

Mr. Webb, Dr. Hugh Dryden (technical director) and Dr. Robert C. Seamans (deputy administrator) asked the writer . . . if guidance for the mission would be feasible during the 1960 decade. The answer was "yes." Asked if the Instrumentation Laboratory would take responsibility for conception, theory, design, test, documentation for production, and assistance into operation for the control, navigation, and guidance system, the answer was again "yes." An inquiry as to when the equipment would be ready was answered by "before you need it." A final question, "How do we know you are telling the truth?" was answered by the writer who said: "I'll go along and run it." This finished the conference.[22]

Draper, who was fifty-nine years old when this meeting took place, later sent a letter to NASA asking to be inducted into the Astronaut Corps, but nothing further came from his offer. Was he serious? Draper was a skilled pilot who loved to test prototype guidance systems by taking the controls of a plane and putting it through its paces. Perhaps he was too old to be an astronaut, but he certainly was serious about the laboratory's commitment to the project. (In 1970 the MIT Instrumentation Lab was renamed the Charles Stark Draper Laboratory in his honor.)

Some sort of inertial-guidance system for Apollo was proposed as early as 1961. But inertial guidance implies computers. In 1961 it was not clear what kind of computers would be required or where they would be located, on board or on the ground. Ground-based computers played crucial roles in all rocket launches, manned and unmanned. Their design and need for Apollo was understood. On board the situation was different. America's first manned space project, Mercury, used no on-board computer whatsoever. Project Gemini, which tested the feasibility of orbital rendezvous, used one and was the first manned spacecraft to operate under some degree of computer control (figure 6.5). The requirements for

6.5

This transistorized computer (shown here with its cover removed) assisted the two astronauts with rendezvous and docking maneuvers of the Gemini space capsule. (Photo: David E. Dempster)

Apollo were hammered out as the Gemini space missions were taking place. Against this backdrop the computing needs for Apollo were apportioned among a complex of on-board and ground systems in meeting after meeting between NASA, MIT, the astronauts, and electronics and computer companies.

Apollo flew with four on-board computers (figure 6.6). Each was a digital, stored-program machine that was part of an inertial-guidance system. The complexity of its mission demanded no fewer, as it included a rendezvous and docking in lunar orbit and critical maneuvers on the far side of the moon, where communication with earth was impossible. One computer was mounted on a ring between the upper stage of the Saturn launch vehicle and the Apollo Modules. This computer, built by IBM, guided the rocket from launch until it boosted the astronauts on a course to the moon.[23] Another was on the Command Module and handled the guidance and navigation through deep space from earth to the moon and back. A third was on the Lunar Module and handled the control of that ship as it descended to the lunar surface. It also handled blasting off from the moon and docking with the Command Module in lunar orbit. Finally, a small computer designed by TRW was part of the Abort Guidance System on the Lunar Module and was designed to get the module and its two-man crew off the lunar surface and back to the Command Module if the primary system failed.[24]

Late in 1963 NASA decided that the two primary computers on the Command and Lunar Modules would use identical hardware, even though they had different missions. That saved development time and gave a small measure of backup in case one failed while the two craft were docked together. (Just such a need arose during the Apollo 13 mission in April 1970 when an explosion left the Command Module virtually without electrical power. The astronauts were able to reprogram the Lunar Module's computer to perform vital guidance and navigation calculations to get them safely back to earth.) Each computer weighed about seventy pounds and was housed in a metal block about two feet by one foot by six inches (figure 6.7).

Astronauts communicated with it by a combination display and keyboard unit called DSKY: the Command Module's computer had two DSKYs, the Lunar Module had one. Each computer also received signals from the gyroscopes and accelerometers of an inertial-guidance system. The Command Module also received signals from an optical system, on which the astronauts took periodic sightings of stars to navigate to and from the moon.

Designing the DSKY was one of the toughest jobs early in the development process. The DSKY is laid out like a modern pocket calculator, except that it is a bit larger to accommodate gloved fingers. In 1962, when it was being designed, there was little agreement that a DSKY was the best way to communicate with a computer. The astronauts, all experienced pilots, were more used to analog dials, meters, and switches. But the flexibility of a general-purpose digital display won out. On the DSKY were three five-digit numeric displays, three two-digit numeric displays, fourteen special purpose status and warning lights, and a keyboard consisting of the ten decimal digits plus nine other keys.[25] Astronauts gave commands to the computer in the form of noun-verb combinations. For example, to display the velocity, an astronaut would key in the two-digit code for "display," then the word "verb," then the two-digit code for "velocity," then the word "noun." Verb and noun codes were displayed on the two-digit displays, while velocity or other data were displayed on the five-digit displays. (Three five-digit displays typically displayed x, y, and z positional coordinates.) It took over ten thousand keystrokes to go to the moon and back.[26]

As was typical of products developed by the Instrumentation Lab, commercial contractors took over the job of manufacturing the guidance system and its computer. The AC Spark Plug Division of General Motors was the prime contractor for the guidance system, and Kollsman Instruments built the optics subsystem. Raytheon Corporation built the computer.

These two guidance and navigation computers used integrated circuits. Eldon Hall, the chief designer at the Instrumentation Laboratory, made that decision in the fall of 1962

6.6 a

Main panel display

Optics: scanning telescope and sextant

Inertial-measurement unit

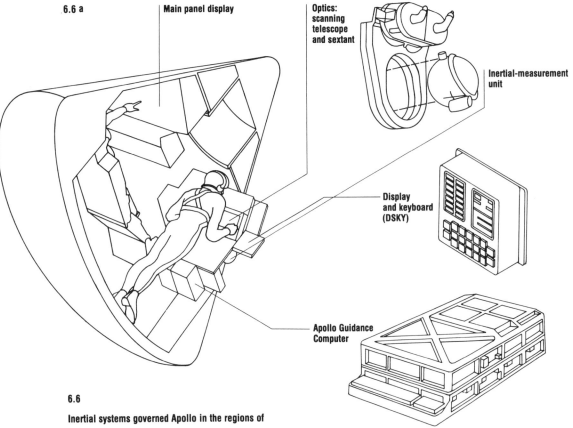

Display and keyboard (DSKY)

Apollo Guidance Computer

First stage

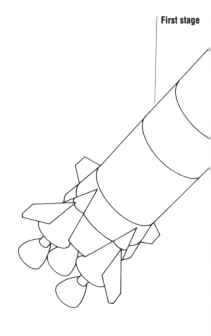

6.6

Inertial systems governed Apollo in the regions of the earth and moon. On the long journey from the one to the other the astronauts drifted along a trajectory determined by Kepler's Laws. They periodically took star sightings, and made minor midcourse corrections to ensure they were on the right course (*a*). The Saturn V–Apollo configuration that carried men to the moon had four general-purpose electronic digital computers on board (*b*). One, on the instrument ring on the upper booster stage, handled the guidance of the rocket boosters until the final boost out of earth orbit. A computer in the Command Module handled guidance, navigation, and control for the entire journey. An identical computer on the Lunar Module handled the descent to the moon, ascent, and rendezvous with the Command Module in lunar orbit. The fourth, the Abort Guidance Computer on the Lunar Module, was intended only as a backup to the main Lunar Module computer in case of a malfunction. (Source: Charles Stark Draper Laboratory)

b

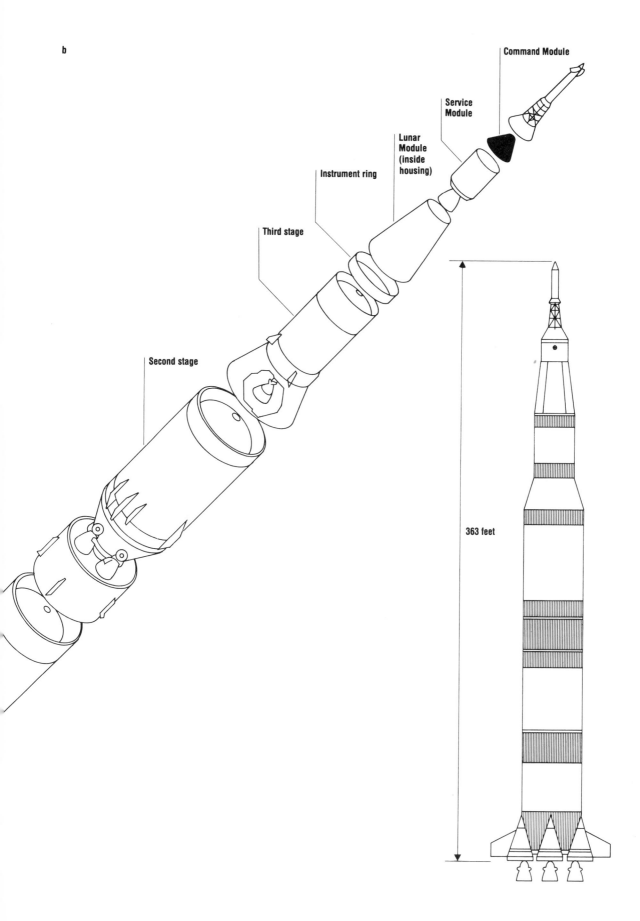

Command Module

Service Module

Lunar Module (inside housing)

Instrument ring

Third stage

Second stage

363 feet

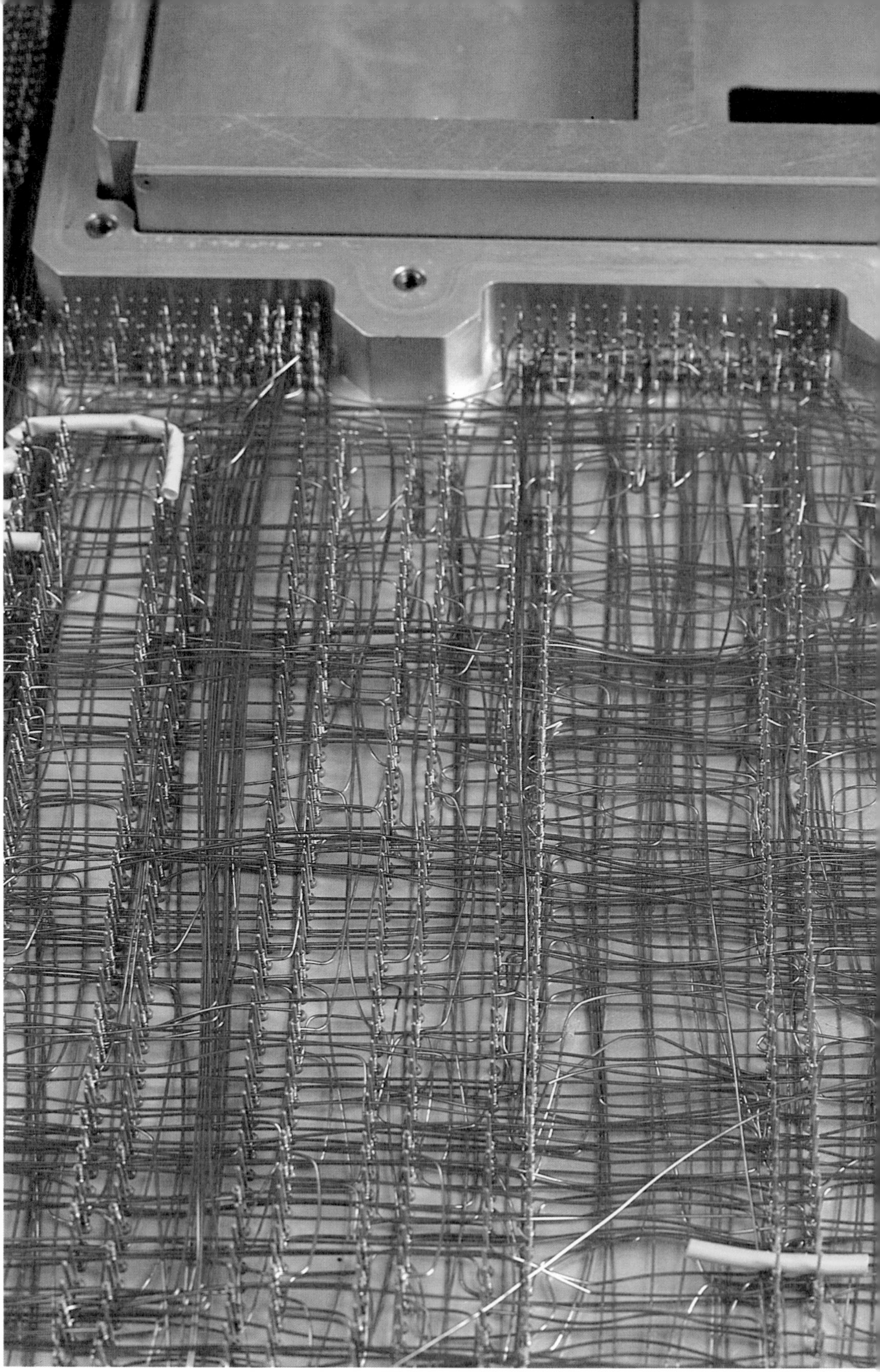

b

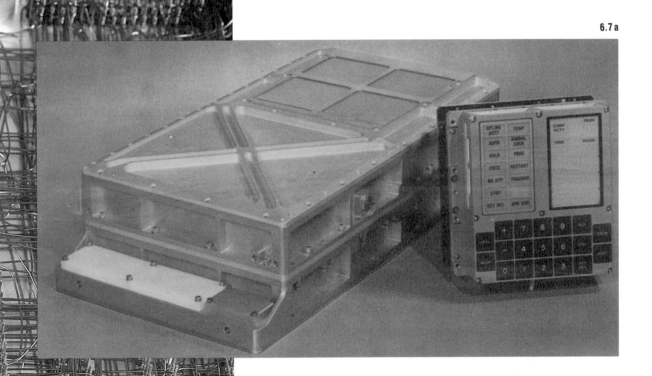

6.7
The Apollo Guidance Computer was a 60-pound,
general-purpose digital computer. The computer
shown here is a prototype used for testing and
programming at the MIT Instrumentation Laboratory (a). Production models of the computer had
their circuits and wiring "potted" in epoxy
resin to protect them from the vibrational stress
of launch (b). (Photos: a, David E. Dempster;
b, Charles Stark Draper Laboratory)

6.8

Eldon Hall was in charge of developing the Apollo
Guidance Computer at the MIT Instrumentation
Laboratory. He chose the then-untested integrated
circuit for use in the Block II Apollo computers,
which were the ones that would be in the craft
that actually went to the moon. (Photo: Charles
Stark Draper Laboratory)

(figure 6.8). He said that the decision came naturally after considering the computing requirements set out by NASA against the extremely limited size and weight allocated on the Command and Lunar Modules for the hardware.[27] In 1962 integrated circuits were only a few years old. Not only did the first specimens Hall got from several suppliers cost more than the discrete circuits they were to replace, they were less reliable. In Draper's words, Hall made "the best sales pitch" Draper had ever heard to convince NASA of his decision, and NASA went along.[28]

MIT designers came up with a design for the Apollo computer that used only one type of chip on which were six transistors plus a few other devices. A complete computer required four thousand chips. Eldon Hall met with both inventors, Jack Kilby of Texas Instruments and Robert Noyce of Fairchild Camera and Instrument Corporation, to inform them of Apollo's needs. After several trips to Texas and California, he chose Noyce's company, Fairchild, to supply the chips for Apollo. After 1963 Philco-Ford became the main supplier of the chips, using Fairchild's design specifications.

Over seventy-five were built, a large number for what many consider an exotic, special-purpose computer. As production mounted the cost of the integrated circuits came down from about $1,000 per chip when discussions with Fairchild began, to $100 per chip for the first Apollo chips, to an eventual $25 per chip.[29] As mentioned earlier, the Apollo and Minuteman development projects consumed nearly all the integrated circuits produced up through 1963.[30]

The Legacy of Apollo

In December 1972 Gene Cernan stepped off the moon, climbed into the Apollo 17 Lunar Module, and with Harrison Schmitt and Ron Evans rocketed back to earth. No one has returned to the moon since, and at present no country has any firm plans to return. As the years since Apollo march by, the wonder and awe over those voyages increases rather than

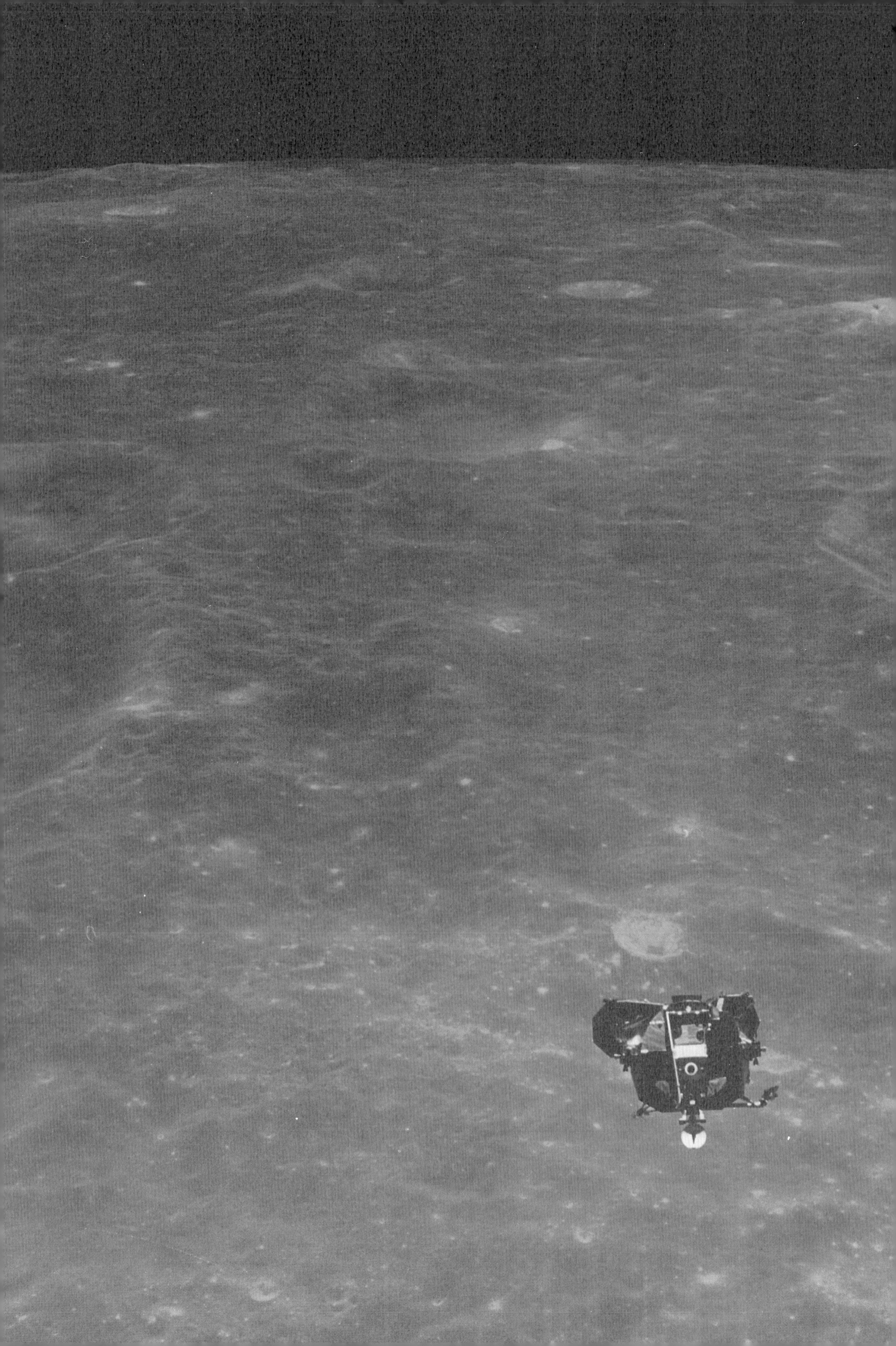

6.9

Images from the Apollo voyages that have been etched onto the world's memory. Astronaut Edwin Aldrin walks on the moon's surface, July 20, 1969 (*a*). Aldrin and Armstrong, in the ascent stage of the Lunar Module, return from the moon to a rendezvous with Mike Collins in lunar orbit (*b*). The astronauts left the descent stage of the Lunar Module, with its engine and spidery legs, on the moon. (Photos: NASA)

6.9 a

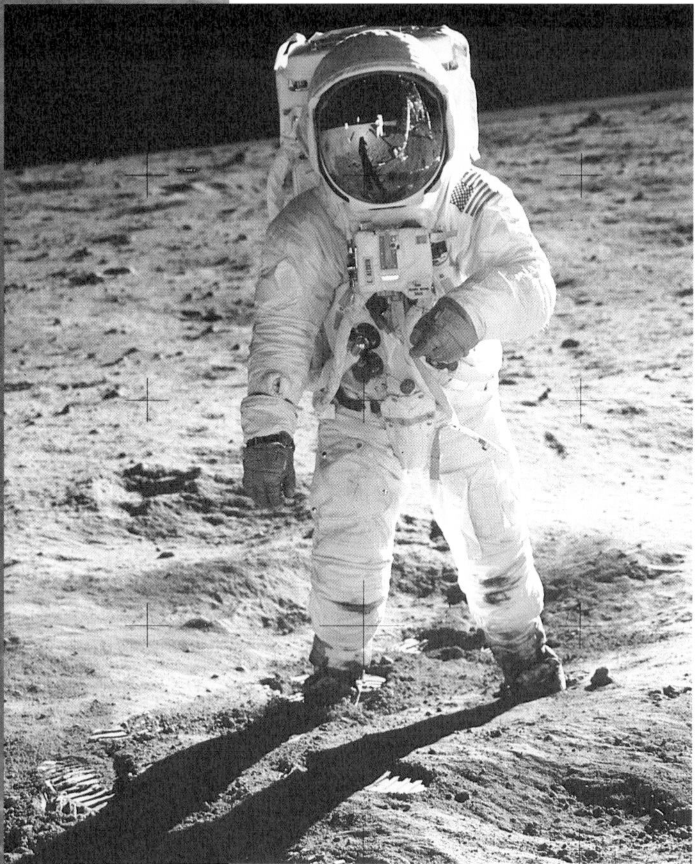

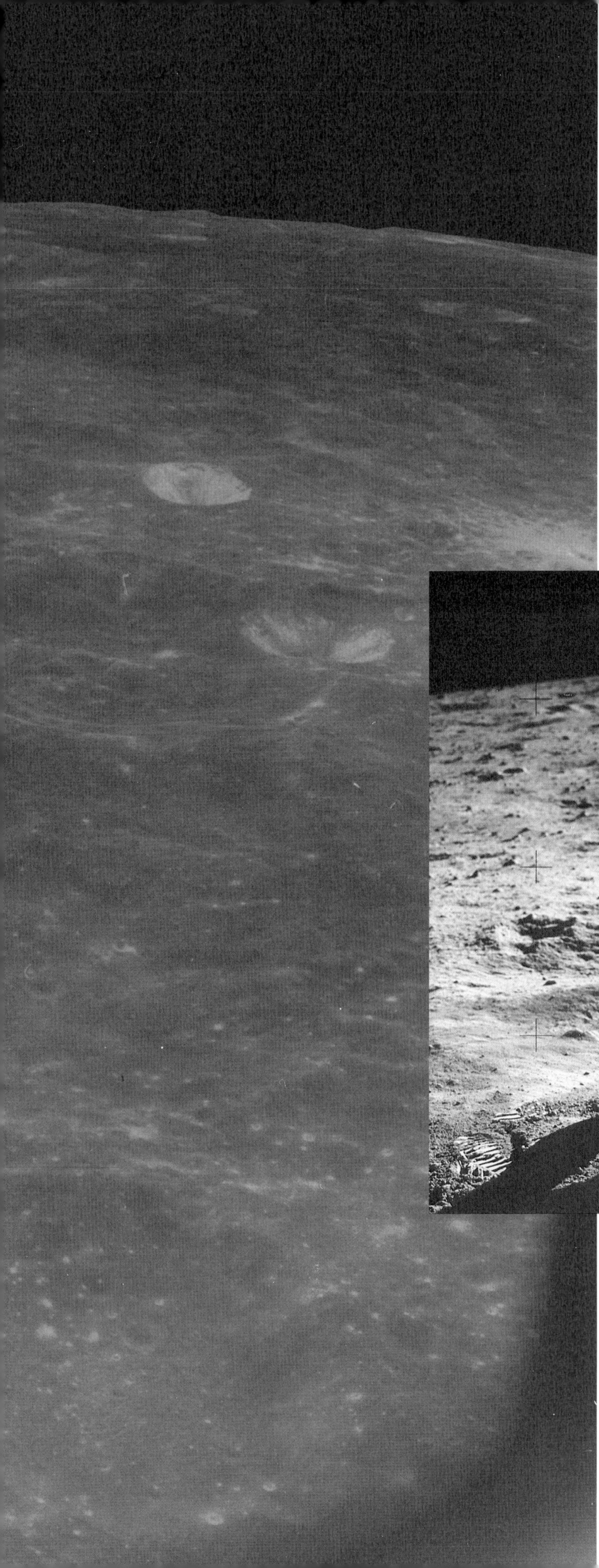

b

fades away. At times it seems hardly possible that it happened at all. The history of the program from President Kennedy's challenge in 1961 to the final mission is one that students of contemporary U.S. culture turn to again and again for a moral or lesson. But the history of Apollo raises far more questions than it answers (figure 6.9).

Some of the most intriguing questions have to do with technology—specifically, the extent to which technology is a driver of society, as some have suggested.[31] Was Apollo an example of technology forcing itself on society, making people do things simply because it was technically possible to do them? Or was Apollo an example of the opposite: technology is not just an application of physical laws but also an expression of social and political needs, an expression of the will of society?

Stated in the extreme, neither view holds true for Apollo or for any modern technical innovation. The history of Apollo, and especially of the development of its guidance and navigation equipment supports both views. Consider first the case for the autonomous power of technology to affect society.

There were many reasons the United States went to the moon. One was that in 1961 engineers recognized that it was technologically possible to go there and they told President Kennedy as much. That was why he picked the moon as the goal and the 1960s as the decade. The rapid strides in inertial guidance and miniaturized digital computing led the engineering community to suggest the moon as a goal. In 1961 no one was confident that the United States could surpass the Soviets in launch technology, but they were confident in the United States's ability to be the first to develop the long-distance navigation techniques needed for a lunar mission. Guidance, navigation, and control were by no means the only major hurdles in the way, but they played a much greater role in a mission to the moon than in any earth-orbiting space venture.

The rapid strides in inertial systems that Doc Draper and the Autonetics engineers made in the 1950s forced changes in political and military strategy elsewhere as well.[32] In the early years of the cold war, the prospect of using nuclear-tipped, long-range ballistic missiles was viewed with skepticism and abhorrence by many military planners. Because of the distances involved and the uncertainties of navigation, warheads had to be absurdly large to have even a remote chance of destroying a target (which meant destroying the city in which the target was located). But as improvements in inertial guidance made those missiles ever more accurate, effective, and precise, resistance to their deployment faded. The present character of the U.S.-Soviet arms race is almost wholly based on the terror of ballistic missiles, against which defense is nearly impossible.

The thousandfold reduction in size and cost of computer circuits, pushed along by Apollo and Minuteman, demanded a rethinking of just about every system that could use computers. We now see that this amounted to rethinking every technical system, large and small. The eventual impact of integrated circuits on American society at large is obvious, as is the extent to which this impact has been propelled by the irresistible temptation of inexpensive computer chips with almost unlimited flexibility.

In other respects Apollo supports the opposite conclusion: that technology is driven (or rather, shaped) by social and political needs. President Kennedy's call for a mission to the moon was a result of cold war politics and a political need to do something spectacular in space. Like the Air Force's need for an air defense system, this call set the pace for raising the level of computer and electronics technologies.

Seen in this light, the commercial development of integrated circuits, with all the wonderful benefits it has brought to society, was not an inevitable and obvious event. Rather it was a product of the social climate of the late 1950s and early 1960s. Without enormous sums of development money from NASA, it would not have happened at all. Recall that in 1961 the integrated circuit was but one of several competing techniques for making electronic components and that the first circuits were neither cheap nor reliable. Apollo, which required huge leaps in the power, reliability, and miniaturization of electronic intelligence, helped bring one of those techniques, the silicon integrated circuit, from the laboratory to a practical commercial product.

The great revolution of recent years has not been in space; rather, it has been in consumer society and in the prosaic business applications of computers. In this view, we owe the present-day consumer electronics revolution to social needs for such devices. Apollo was not the driver but the catalyst of these events. The need existed independently of that project and of the technical issues of how cheap and small circuits could be built.

However one looks at these events, between 1960 and 1975 computers got dramatically smaller, faster, more reliable, and cheaper. Scientific Data Systems, Digital Equipment Corporation, and Data General introduced a new type of computer, the mini, on the market. Owing mainly to their use of integrated circuits, these machines were small, rugged, fast, and inexpensive. Immediately after their introduction they began selling in unprecedented volumes. (The PDP-8, from DEC, has been called the first modern minicomputer because of its unique combination of small size, low price, and high performance. It was first delivered in 1965.)[33]

In 1971 Ted Hoff, an engineer at Intel Corporation, put a variant of the computer architecture of the PDP-8 on a single silicon chip. The idea for doing that came from a request by a Japanese calculator manufacturer to build a set of chips for its line of pocket calculators. Hoff responded by designing not a whole set but just a single chip, the microprocessor, which could be programmed to do different things like a computer. It was thus versatile enough to serve the entire calculator line, which saved money and time (and helped retire the slide rule). By having a complete processor on a single chip, an engineer could easily design intelligence into other machines. This transformation of the machines that surround us is the most important part of the story about the development of technology in the past fifteen years.

All the pieces of this picture fell into place by about 1975, ironically just as the last Saturn rocket boosted the last Apollo astronauts into space for an earth-orbit rendezvous with two Soviet cosmonauts. Aerospace, like so much of the rest of society, felt the profound effect of the silicon chip. The computer has since transformed nearly every aspect of air and space flight, and it continues to do so. The next two chapters outline some of the things that have been happening.

Chapter 7
Advances in Design and Production

By 1975 the level of computing technology reached a height sufficient to cross the aerospace threshold in a big way. The integrated circuit was of course the basis for much of that attainment, but there were other factors at work as well. What follows is a brief look at the state of computing in that year.

By 1975 the integrated circuit was the basic building block of computers. That made for small, lightweight equipment with low power requirements. It was now easier to design and build digital computers for on-board applications. And back on the ground a typical computer installation became far more powerful. Not only could one now put the power of a 1950s-vintage computer in one's shirt pocket; one could also fill the computer room in a university or plant with machines that were thousands of times more powerful. While on-board computers would have a great impact on flight navigation and control, these other, ground-based machines would have a similar effect on aerospace engineering and design and on the science of aerodynamics as a whole.

The use of solid-state circuits, aided by a strong push from programs like Minuteman and Apollo, helped solve the reliability problems of the vacuum-tube era. This too hastened the spread of digital computing into air and space flight, especially where human lives were at stake.

Advances in circuit design and memory technology led to computers that operated much faster than they did in the 1950s. By 1975 processing speeds had advanced to a point where digital computers could take on the real-time applications, such as flight control and rocket guidance, that previously were the domain of mechanical analog devices.

Finally, advances in software and programming techniques enabled aerospace engineers, pilots, scientists, and astronauts to make better use of computing power. The Apollo Guidance Computer inaugurated the method of having a pilot communicate directly with a computer through an interactive keyboard and display. SAGE pioneered graphically rich display screens in place of indecipherable printouts of numbers. Other programs, such as the RAND Corporation's JOSS, developed for its own employees, permitted users to communicate with a computer in simple, easy-to-remember commands. These innovations helped move the machine out of the domain of computer specialists.

Not all of these innovations were direct results of aerospace demands. But no matter, they penetrated into every flight activity. By 1985 this process reached a level at which digital computers were thought as much a part of flying as wings and engines. For each flight activity, however, the effect was different. The following sections are not an exhaustive treatment of the subject but seek instead to describe some of the more dramatic examples of the impact of the computer. This chapter describes the changing nature of the design of air and space craft; the next chapter describes the effect of the computer on flight operations and pilot training.

CAD/CAM

Today one of the most common terms heard in the computer and aerospace industry is "CAD/CAM," which stands for "computer-aided design/computer-assisted manufacturing." The popularity of the term reflects the central role design has in aerospace. The first step in the long journey from concept to flight is design. It is the process by which ideas are turned into airplanes, rockets, and spacecraft. Design is the key to understanding aerospace technology. The historian of technology Edwin Layton has said of it: "Design is an adaptation of means to some preconceived end. This I take to be the central purpose of technology. . . . We may view technology as a spectrum, with ideas at one end and techniques and things at the other, with design as a middle term."[1] To the extent that computers assist with design, they will have penetrated to the very core of aerospace activity.

For any engineering discipline, but especially for aerospace, a successful design involves numerous compromises and juggling conflicting demands. A paradox of good engineering is that one obtains a better product when one abandons a search for absolute perfection. Knowing when to compromise and when to hold one's ground is what distinguishes the good engineer from the crowd.

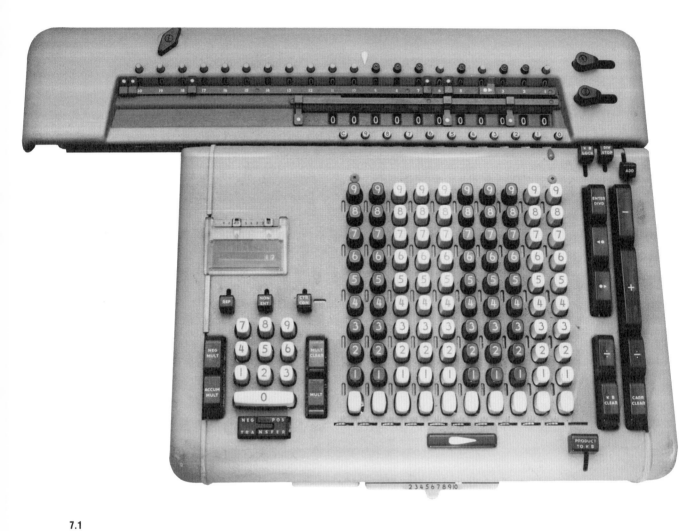

7.1

This Friden calculator was the mainstay of numeric calculations before the advent of computers. (Photo: National Air and Space Museum)

A successful aircraft or rocket design strikes just the right balance among mutually conflicting demands for safety, performance, payload capacity, fuel consumption, ease of manufacture, and cost. To achieve that balance, a designer typically proceeds through a number of cycles of the design process. That process goes from preliminary design to analysis to a final set of detailed specifications. With luck, this process converges to a good design; its result is a final set of detailed drawings and other information that is turned over to production engineers, who build the craft according to the design team's wishes. (Turning over the detailed drawings to the production group does not end the process of design; shop foremen and others on the factory floor can and usually do suggest design changes that reflect what is feasible to produce at reasonable cost and under given time constraints.)

Computation plays a role in each of these phases of the design cycle, but for each phase the role is different. The fact that a design goes through this cycle many times means that vast amounts of data have to be stored and retrieved under a rigid system that ensures that when a design is changed, all the relevant files are updated and properly documented. The standard way of implementing changes is through a formal notice called an engineering change order, which is approved by the team leader (or a supervisory group) and then distributed to everyone who might be affected by that change.[2] It is not unusual for an airplane design to undergo *thousands* of engineering change orders during its design. Clearly then, mechanical aids to tracking and recording these changes are a necessary part of design.

Another obvious role for computers is to assist in the second phase, structural analysis (see chapter 2). Structural engineers, using their knowledge of the properties of materials, look at the preliminary design and determine how strong each piece must be to bear loads with an acceptable margin of safety. Before the advent of digital computers, the stress analyst's tools were pencils, paper, and a mechanical desk calculator (the Friden was a favorite [figure 7.1]). With the introduction of electronic computers like the IBM 701 in the

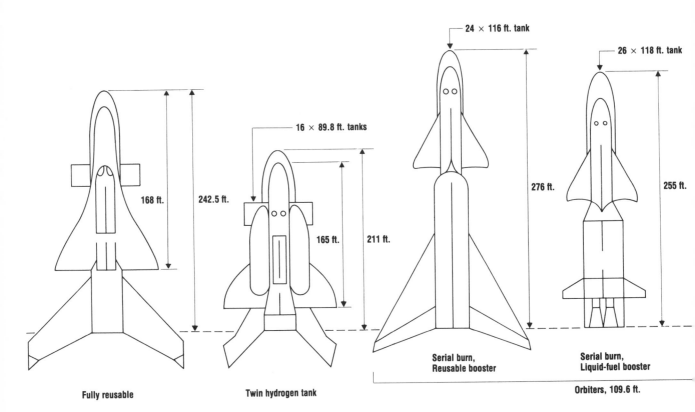

24 × 116 ft. tank

26 × 118 ft. tank

16 × 89.8 ft. tanks

168 ft.   242.5 ft.

276 ft.   255 ft.

165 ft.   211 ft.

Fully reusable

Twin hydrogen tank

Serial burn,
Reusable booster

Serial burn,
Liquid-fuel booster

Orbiters, 109.6 ft.

7.2

The Space Shuttle went through a large number of
initial configurations from the 1960s to March
1972, when the present configuration was agreed
upon. (Source: John Logsdon, The Space Shuttle
Program: A Policy Failure? *Science* 232 [May 30,
1986], p. 1100)

early 1950s, the Friden gave way to the central mainframe. Engineers solved equations by submitting tall stacks of punched cards containing their Fortran programs. This work, tedious, plodding, but necessary, is still the backbone of aerospace design today. Most aerospace companies have a central mainframe computing facility to handle these jobs, supplemented by a clutch of powerful minicomputers, like Digital Equipment Corporation's VAX, to handle detailed computations for smaller engineering teams.

What transformed design in the 1970s was the computer's ability to take over some of the work of the first and third phases of the design cycle. The medium in these phases is not numbers but drawings. In the preliminary phase the designer tries out a number of different configurations in search of an optimal one (figure 7.2). In the final phase rough sketches have to be turned into detailed drawings in which every last rivet and piece of metal is specified and drawn in. When computers appeared with that ability (or at least some of that ability), the era of CAD/CAM began.

It is not enough to have the computer display information in graphic form, as the SAGE system did. CAD/CAM requires that the machine manipulate and store pictures, just as it handles numbers. It must handle not just the picture but also the numeric information associated with each picture. If a computer has a drawing of a wing spar, for example, it must also store and make available, when needed, the dimensions that go with it. If the designer alters something, not only the picture but also the associated data must change to reflect that alteration, and these changes must be reflected in the engineering change order that the designer sends out. Finally, there has to be a way for the engineer to draw, reshape, and erase lines naturally, as one would with more traditional tools.

In 1961 Ivan Sutherland, a doctoral candidate in electrical engineering at the Massachusetts Institute of Technology, began work on a system he called Sketchpad, which he hoped could do all these things (figure 7.3). His program was written for MIT's TX-2 computer, one of the first transistorized machines and one with sufficient memory to handle the demands of

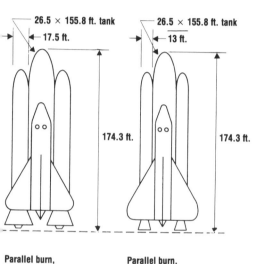

26.5 × 155.8 ft. tank

17.5 ft.

174.3 ft.

**Parallel burn,
Liquid-fuel booster**

26.5 × 155.8 ft. tank

13 ft.

174.3 ft.

**Parallel burn,
Solid-fuel booster**

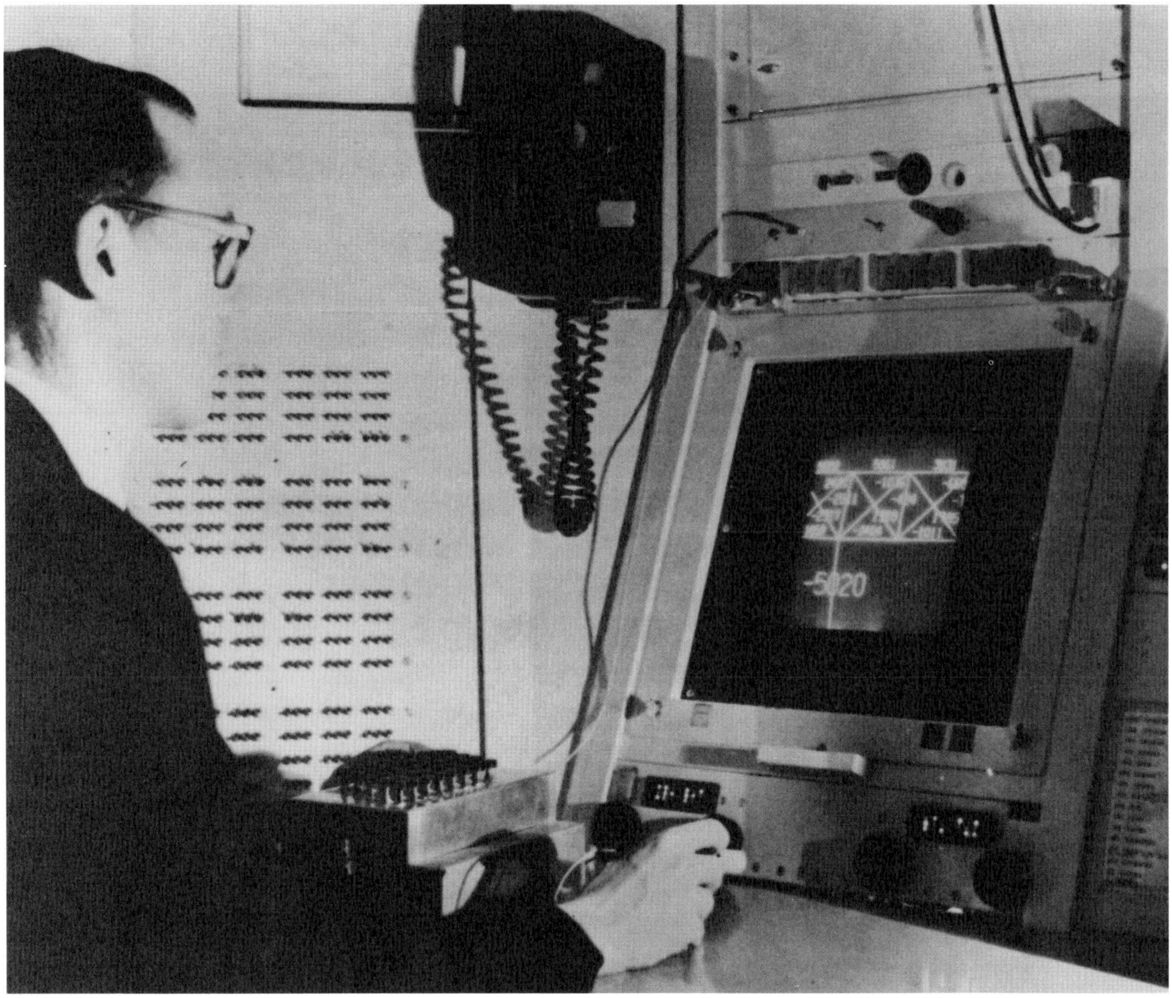

7.3 a

b

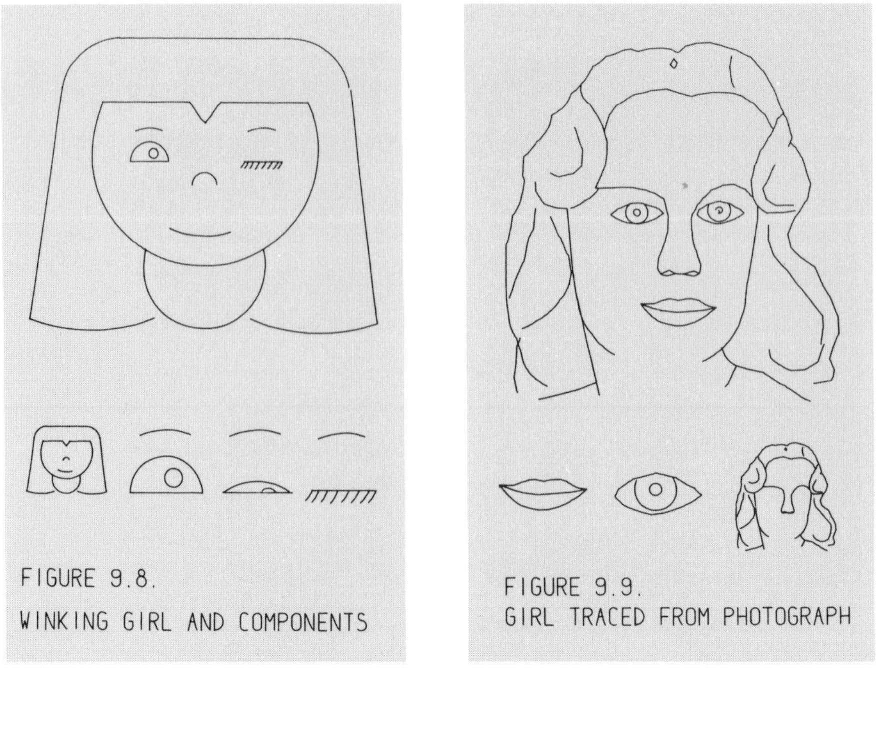
FIGURE 9.8.
WINKING GIRL AND COMPONENTS
FIGURE 9.9.
GIRL TRACED FROM PHOTOGRAPH

graphics processing. Two years later Sketchpad had progressed to a point where it could compete head-to-head with a professional draftsman using standard drawing tools. By 1964 most of the big computer, automobile, and aircraft companies were building systems like Sketchpad. IBM's introduction of a general-purpose graphics terminal for their 360 series of computers gave a measure of respectability to the whole enterprise. The IBM model 2250 Terminal was expensive, and few were sold; nonetheless, it was a start.

In 1968 Ivan Sutherland joined David Evans of the University of California's Computer Science Department to form a company devoted to computer graphics. They introduced a terminal that was much more versatile than IBM's, but at twice the price ($250,000) it could hardly begin to displace human draftsmen. The price breakthrough finally came from a small electronics firm in Oregon called Tektronix, which developed a special video tube that could retain an image on its surface without tying up a large and expensive chunk of the computer's main memory. By the early 1970s one could purchase a system based on Tektronix's Direct View Storage Tube for $20,000, a price that made computer-aided design far more attractive. Ten years later these companies introduced a further innovation, displaying crisp color drawings on a raster screen similar to those used in ordinary television sets. This innovation allowed greater versatility while keeping costs down. With these high-resolution raster displays, CAD/CAM finally became practical.

7.3

Ivan Sutherland is here seated before the TX-2 computer at Lincoln Laboratory (*a*). On the screen is a Sketchpad depiction of a bridge truss. Some Sketchpad drawings from Sutherland's doctoral dissertation (*b*). (Photos: Evans and Sutherland Corporation, National Air and Space Museum)

CAD/CAM in Action

Let me give a condensed walk-through of the design of a new airplane, say a medium-range commercial passenger jet.[3] In its initial phase of preliminary design, a team of perhaps twelve people study dozens of designs to come up with about five or six configurations. These they turn over to another team of perhaps a hundred people, who take these five or six options and come up with an optimal design. More and more people are drawn into the process as this optimal design takes form, and in particular, as each of the subsystems (e.g., the landing gear) are specified in greater and greater detail.

FIGURE 9.10.
GIRL WITH FEATURES CHANGED

The preliminary-design group immediately has to face some tough decisions. Before the design process starts a number of parameters are already fixed. For example, if the airplane is intended for the U.S. domestic market, customers will surely want to use it at the New York LaGuardia Airport, one of the busiest in the world. All of the runways at LaGuardia are fairly short, and one is on pilings over a bay. These conditions place immediate restrictions on the take-off weight and landing gear parameters of the aircraft. Because the airport is surrounded by residential neighborhoods, there are stringent noise regulations in force, which a plane must meet by utilizing quiet engines, a high rate of climb, or both.

But other choices are not as straightforward. How many engines: two, three, or four? Where should the engines be mounted: on pylons under the wing or at the rear of the fuselage? Should the wings be mounted high or low on the body? Should the rear stabilizer be mounted on the fuselage in line with the wings or high on the rudder in a T configuration (figure 7.4)? There are benefits and drawbacks for each choice. Consider the following factors, which are only a small subset of the many trade-offs implied by these decisions.[4]

Mounting the engines on pylons under the wing makes servicing them easier, but pylons add weight, and the plane needs a longer (and therefore heavier) landing gear to make sure the engines clear the ground. Rear-mounted engines avoid these problems and allow the wing to be aerodynamically cleaner, but mounting the third engine inside the fuselage creates maintenance problems. Engine placement also affects the plane's center of gravity, which may necessitate a change in the placement or shape of other parts of the craft. Fewer engines mean lower maintenance costs, but the engines have to be bigger, and therefore heavier and more complex, to produce the equivalent thrust. That might nullify the fuel and maintenance savings the designer hopes to achieve. A high wing gets pylon-mounted engines well away from the ground and allows a short landing gear. But it also means that the landing gear has to be mounted in the fuselage, and this makes the wheelbase narrower and therefore less stable. A T-shaped tail has aerodynamic advantages, but the tail fin has to be made stronger and therefore heavier to bear the stress.

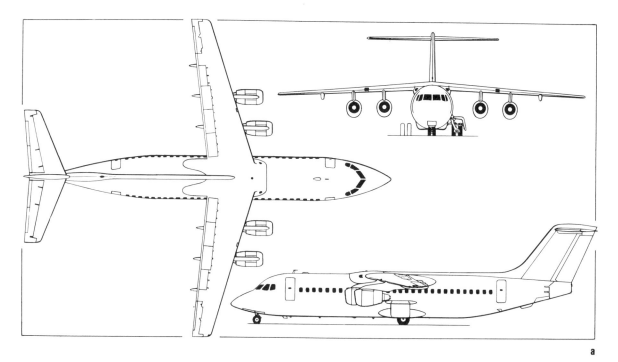

a

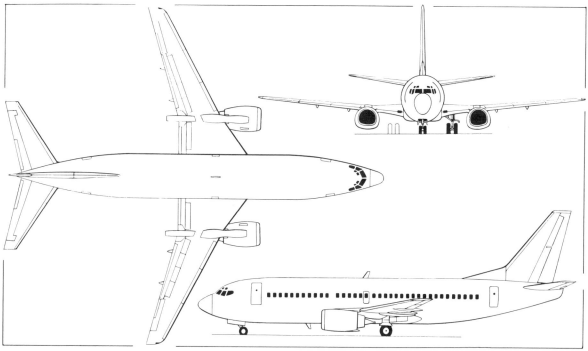

b

7.4

Two successful commercial passenger jets, the British Aerospace 146-300 (*a*), and the Boeing 737-300 (*b*). Both serve the same markets yet represent very different design philosophies. Note the placement and number of engines, the location of the wings on the fuselage, and the tail configuration. (Photos: Pilot Press)

There is no general rule of thumb and certainly no computer program to tell a designer which of these options to choose. It would be hard to find a more challenging task in any other branch of engineering. Even with all the computing power of computer-aided design, design remains a matter of human judgment and skill.

The medium of communication during preliminary design is still sets of drawings. Drawings, including full-sized drawings on large sheets of paper, remain the preferred way to tell others in the plant what the design is like (figure 7.5). Producing these drawings is called lofting, a term borrowed from shipbuilding, where hull designs are laid out in large lofts. (Some of the same rooms were taken over by aircraft companies for their drawing needs.) By the end of this phase a thousand or more employees will be working on the production design, with the eventual goal of producing new sets of detailed drawings that the manufacturing arm of the company can use.

It is in the production of these drawings that CAD/CAM has enjoyed its greatest success. To go from an initial sketch to a layout drawing requires adding enormous quantities of detail to fill in the unknown with rivets, flaps, screws, struts, windows, wires, and plumbing. For many standard pieces, such as cockpit instruments or landing gear wheels, the designer can call up a complete and detailed drawing from a standard library of shapes stored in the computer's memory. If what is in the library satisfies the requirements of the new design— and often that is not the case—using these data can save a lot of time.

This design activity has to dovetail with the work of the stress-analysis people, who establish the structural integrity of the proposed design. With the advent of CAD/CAM workstations this work too has become more interactive. Advances in computer processing speeds and memory have made feasible a previously impractical technique, so-called finite-element analysis. With this technique the strength of a complex structure is calculated from the stresses on each load-bearing member of a structure. For this technique to work for a typical, complex design, one has to analyze a structure in terms of thousands of its individual load-bearing members, and calculate the stresses at the places where they join one another. Without the digital computer the technique would be impractical, but it is now a routine part of the stress-analysis phase of design.[5]

The design of the critical shapes of wings, rudders, and other aerodynamic surfaces is only a small part of the total design effort, but it is important to the commercial success of the product. No standard library data exists for these shapes, as their performance must be precisely tailored to the particulars of the design.[6] Here is where the computer's ability to calculate comes into play. What the designer can do is to specify cross sections of these critical shapes and let the computer generate the complex curves of the wing surface that smoothly connect one section to another.

Early airplanes looked like boxes with flat plates attached to them. They were easy to design but did not fly very well, as the sharp edges and corners set up air turbulence and friction, which severely impaired performance. A good design should be smooth: wings should blend cleanly into the fuselage, the engine housings should fit well with the wing structures, and so forth (figure 7.6). There are no simple rules for designing these shapes; designers must instead use their best judgement combined with a knowledge of aerodynamics to make the transitions.

A draftsman's tool kit contains triangles, straightedges, compasses, and templates for drawing circles and ellipses. It also contains templates for drawing French curves, curves with a changing radius. Another tool is called a spline, a flexible rule that can draw smooth curves from one straight line to another. These last two tools can help draw the curves needed to produce smooth transitions between various parts of an airplane. Splines can also be generated by algebraic equations, by generating the equations for the set of curves of different

7.5
Laying out full-scale drawings for aircraft wings, a
process called lofting, requires huge rooms. In
some cases aircraft companies purchased or
rented lofts from shipbuilders, where in an earlier
day ships' sails and hulls were laid out. (Photo:
National Air and Space Museum)

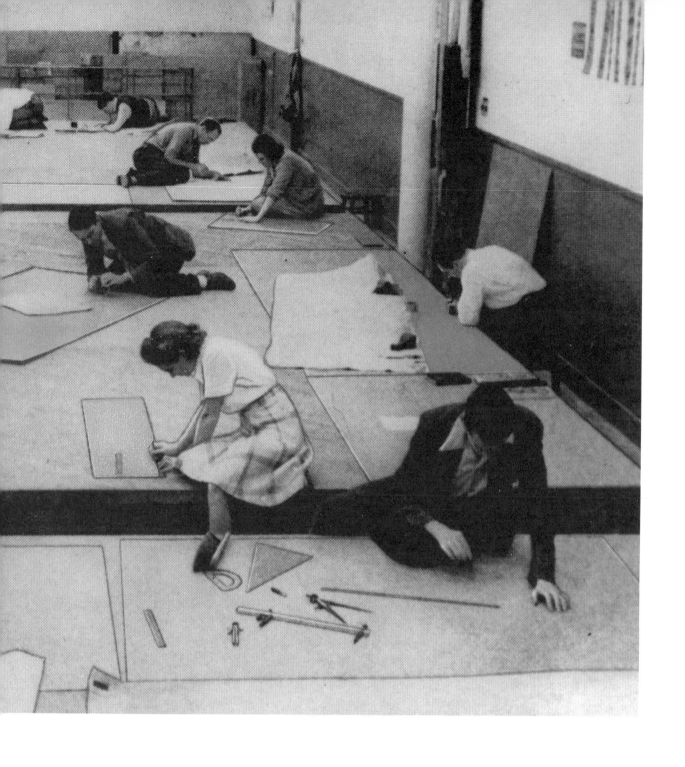

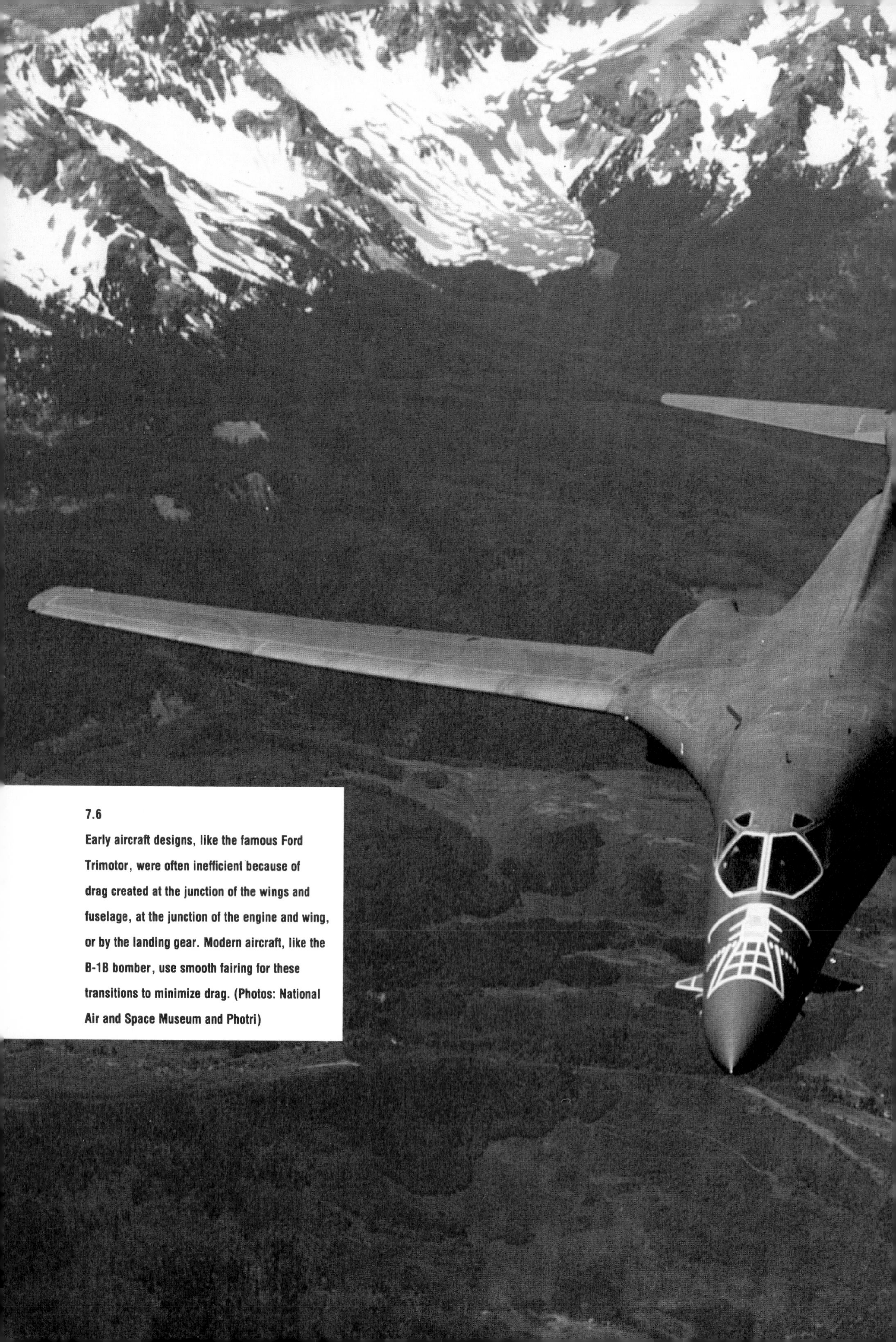

7.6

Early aircraft designs, like the famous Ford Trimotor, were often inefficient because of drag created at the junction of the wings and fuselage, at the junction of the engine and wing, or by the landing gear. Modern aircraft, like the B-1B bomber, use smooth fairing for these transitions to minimize drag. (Photos: National Air and Space Museum and Photri)

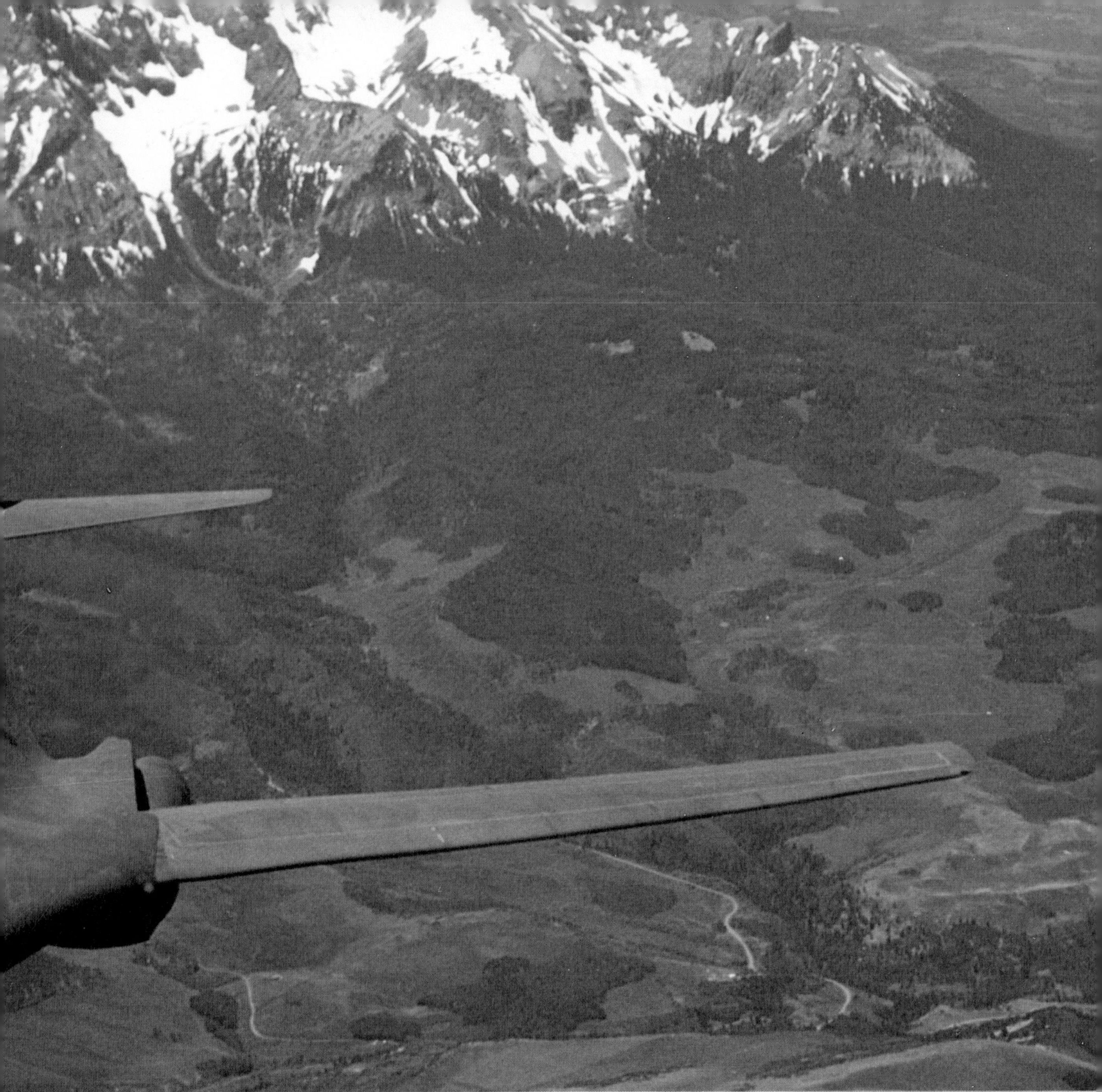

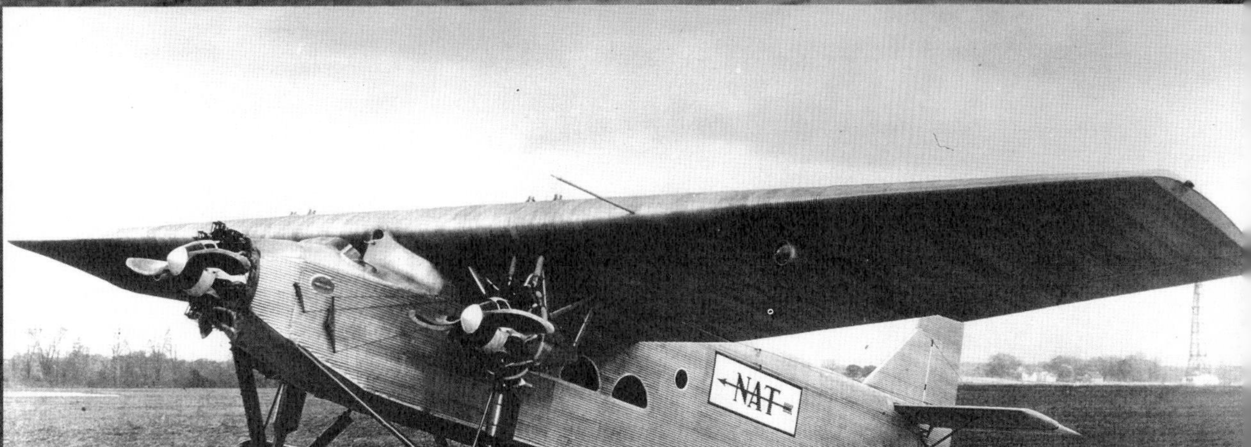
NAT

radii that make a smooth transition from one surface to another. The mathematics required to generate splines typically involves higher-order equations, which are very hard for humans to evaluate by hand but are part of the repertoire of a CAD/CAM workstation.[7]

Generating splines is a job that marries the computer's newly developed ability to draw to its traditional job of "number crunching," for which it was invented in the first place. Generating splines was one of the first applications of computers that competed with manual methods. Computers were first used for this application in the mid-1950s, though at that time the techniques for displaying the curves on a terminal were not yet available. Computer-generated splines result in transitions that are not only smooth but structurally sound, aerodynamically clean, and easy to manufacture.

Manufacturing

So far all the discussion has been on the CAD side of CAD/CAM. What about the computer's role in the manufacturing process—has there been a corresponding transformation of work on the factory floor?

The answer to this question is mixed. In the manufacturing of air and space craft, numerically controlled machine tools are routinely used and are indispensable to the accurate machining and close tolerances that flight hardware requires. But progress here has been more incremental than revolutionary. Most manufacturing still depends on the skill of the human eye and hand in shaping materials. In fact, one result of early efforts at computer-assisted manufacturing was entirely unanticipated: the difficulties that arose when people tried to automate the work on the factory floor has increased our awareness of just how critical the skills of the human machinist are. We understand now in ways we did not understand before that the machinist's hands and eyes contain as much engineering knowledge as the design computers of the structural designers or aerodynamicists.[8]

In discussing computer-assisted manufacturing the term "robot" invariably comes up. It is a term that most people vaguely understand, but for which there is no agreed-upon definition. One essential part of every computer system is a set of input and output devices that allow the computer to interact with the rest of the world. Computers normally interact through the printed word or number. In CAD/CAM this mode of interaction is supplemented with pictures. If one connects a computer's output to a motor, and its input to sensors that measure light, temperature, position, and so forth, one has a robot.[9] It need not look like a human or even be able to move autonomously (although many robots can do so). With the proper programs such a device can, in theory, do the work of a machinist or other worker on the factory floor.

The notion of having a computer assist in the production of aircraft seems a natural extension of the high level of mechanization of mass production that began with Henry Ford's Model T assembly line. But that is not the case, mainly because of the differences between aircraft and automobile production but also because of the inherent capabilities and limits of computers themselves. The auto industry has made great strides in using robots to handle many of the repetitive, tedious, and physically dangerous tasks of assembly, like spray-painting the body, for example. But in aerospace such straightforward substitutions of robot for human are less frequent.

One reason is that it is rare to build an aircraft in long, sequential production runs. A sustained output of hundreds of units per day, routine in any automobile plant, is unheard of in the aircraft industry, even for the most successful craft like the Boeing 737 transport.[10] And for spacecraft and rockets it is even more unheard of.

A second factor is that a much higher proportion of air and space craft consists of parts that must be machined to high precision and close tolerances. Many parts of an automobile, except for the engine and drive train, are deliberately designed to loose tolerances to keep costs down. Automobile production engineers often go to great lengths to save a few *pennies* per car in production. But the trade-offs for cost are different for aircraft. For something like a deep-space probe there is little room for any cost saving that might reduce operational lifetime.

Finally, in the initial phases of production, air and space craft designs frequently change from one model to the next. (For example, there are many differences in the designs of the Space Shuttle orbital vehicles, of which five were built, including the *Enterprise*.) The difference between a successful and a poorly performing airplane may be a matter of a few pounds of weight or a few miles per hour at top speed. There is therefore a great incentive to incorporate immediately any improvements gleaned from the experience of flying the initial versions of the craft.[11]

Therefore, only in the area of high precision can robotics make a difference in aerospace manufacturing. As with so much else in computers, the potential for controlling machine tools automatically and with high precision was foreseen by a number of people a few short years after the invention of the digital computer in the late 1940s.

The present state of the art derives from a proposal in 1949 by John T. Parsons of Parsons Corporation in Traverse City, Michigan, to the U.S. Air Force to use milling machines controlled by punched tape to manufacture helicopter blades. Parson's initial idea was taken up at the MIT Servomechanisms Laboratory, where what is now known as numerical control took form with Air Force support.[12] At the Massachusetts Institute of Technology Douglas T. Ross (who had come to MIT in the fall of 1951) took the next step of developing a specialized programming language that allowed the production engineer to specify what he wanted in terms the computer could understand.[13] MIT introduced a numerically controlled milling machine in 1952. In 1956 Ross joined in a proposal to the Air Force to develop an advanced numerically controlled system capable of specifying three-dimensional parts and controlling three- and five-axis milling machines. Ross demonstrated the preliminary results of the group's work in February 1959.[14]

In brief, a numerically controlled system works as follows: A person enters geometrical information into a digital computer from a set of detailed drawings. Commands are written in the APT programming language. The computer takes these commands and produces a reel of punched tape. This tape is then fed into a milling machine, which then automatically selects the tools and cuts the metal to the desired shape.

APT caught on quickly; it had just the kind of capability to meet the machining demands of jet-powered supersonic aircraft then going into production. By the mid-1960s APT runs accounted for a third of the work then being done on the large computer installations at aircraft plants.[15]

Nonetheless, the degree to which APT has automated manufacturing is far below the degree to which computer graphics have changed design. And there is still no effective link between CAD and CAM, despite the general feeling that such a link is both desirable and technically feasible. The main reason is that the computer software that describes and generates lines and shapes on the designer's screen is nothing like the language that describes the movements of a milling machine to produce that same curved shape. At present it is not technically feasible to try to have a computer program translate CAD output to CAM input.

What might work is a new, common language that both camps can use, but such a language is not easy to produce. The very success of APT and the present CAD systems stands in the way of a common language. Standard protocols for the transfer of design information are arriving, but only slowly. Computer, automobile, and aircraft companies have joined in supporting MAP (Manufacturing Automation Protocol), a set of standards devised by Boeing and General Motors that allows the user to transfer production specifications across computers from different vendors. Also being proposed are standards for describing geometric shapes, such as IGES (Initial Graphics Exchange Standard). But these alone do not integrate design with manufacturing. At present, computer-integrated manufacturing (CIM) remains little more than an acronym in the babel of computer programming systems.[16] In the meantime, engineering drawings will remain the medium by which design information is passed to manufacturing.

Aerodynamics

One part of aircraft design deserves special mention. That is aerodynamics, the study of the forces that air exerts on a solid body moving through it. Aerodynamics is the essence of flight; it is what makes an airplane fly. It is an integral part of the design process: not only the design of the wings; but also the design of the vertical and horizontal stabilizers (the empennage); the design of the flaps, slats, ailerons, and other control surfaces; and the design of the fuselage as well. Aerodynamics is an important part of spacecraft design as well: rockets must be designed to withstand enormous, fast-changing aerodynamic forces on their way through the atmosphere into space, and manned spacecraft like the Space Shuttle must do likewise on their return to earth. Even craft designed to spend their entire existence in outer space, like the planned U.S. Space Station, are subject to aerodynamic forces from the thin atmosphere of a low earth orbit, and these must be understood and analyzed. (Indeed, unforeseen aerodynamic forces brought the U.S. Skylab space station into a premature fiery reentry in 1979.)

The science of air in motion is a branch of theoretical physics and is part of a more general field known as fluid dynamics. The subject has attracted some of the greatest men and women of science. Isaac Newton laid the theoretical foundations in the seventeenth century, but for two centuries after Newton there was little progress. The present understanding of fluid flow is far from complete. Many other branches of physics for which Newton laid theoretical foundations (e.g., optics) are now well understood. But that is not true of aerodynamics.[17]

Navier and Stokes

The reason aerodynamics was not well understood has to do with the mathematics of fluid flow. Newton made the first attempt at a mathematical analysis in the seventeenth century. In 1827 Claude Louis M. H. Navier, a professor at the École Polytechnique in Paris, derived for the first time a set of equations describing fluid motion. In 1845 these were refined and elaborated by the British mathematician Sir George Stokes.[18]

These equations of fluid motion took the form of five second-order partial differential equations, and when expressed in Cartesian coordinates (the familiar x, y, and z axes) they contain over sixty partial derivatives.

A solution of Navier-Stokes equations would in theory give the values of the forces of lift and drag acting on a wing moving through the air at a given velocity. Yet a direct solution is beyond the ability of current mathematical techniques for all but the simplest cases. The Navier-Stokes equations, powerful though they may be, are therefore of little use to an aerodynamicist concerned with analyzing the flow of air over a complex wing form with its flaps, engine nacelles, rivets, and so forth.

For the last one hundred years the chief tool of aerodynamics research has thus been the wind tunnel. Wind tunnels come in all sizes and shapes: some of them fit on a desk-top, others are among the largest man-made structures on earth (figure 7.7). They all operate on the principle that one can construct a physical model of the phenomenon one wants to investigate, take measurements from the model, and apply the results to actual air or space craft, using appropriate techniques to extrapolate and scale up the model data.

Wind tunnels have served aeronautics well since the beginning of powered flight. The Wright Brothers built a small tunnel to test a variety of wing shapes in 1901 and 1902; these tests were crucial to their successful flight the next year. The most difficult problem in using a wind tunnel is that one can never scale up tunnel data without introducing some error or distortion.[19] Tunnel designers have come up with all sorts of ingenious ways to overcome this problem, such as pressurizing or cooling the air in the tunnel, using helium instead of air as the working fluid, and other techniques. They have also developed a theory that tells them in advance precisely how scale factors affect measurements made on models.[20]

These techniques have just barely kept up with the increasing demands of the aircraft designer. Aircraft are being designed to go faster and higher. The Space Shuttle is an extreme case,

7.7

The American inventor Hiram Maxim built this Whirling Arm in 1890 to test the efficiency of propeller designs (*a*). Whirling arms were soon eclipsed by wind tunnels as the principal tool of the aerodynamicist. By the 1970s wind tunnels had grown both in size and sophistication. The tunnel shown here is located at the NASA Ames Research Center in Mountain View, California (*b*). It is big enough to house full-sized aircraft in its test section. (Photos: North Wind Picture Archive and NASA)

b

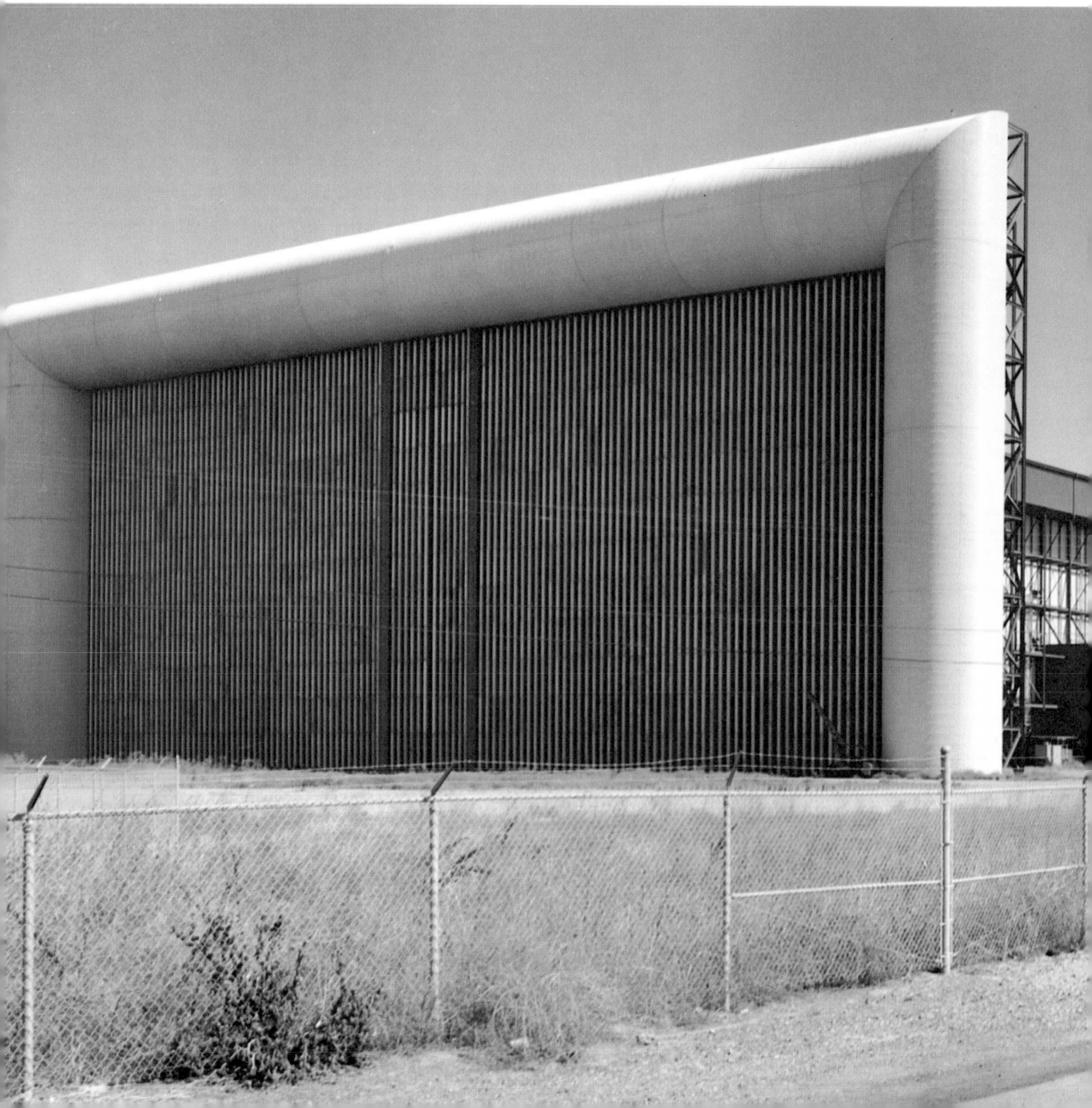

a

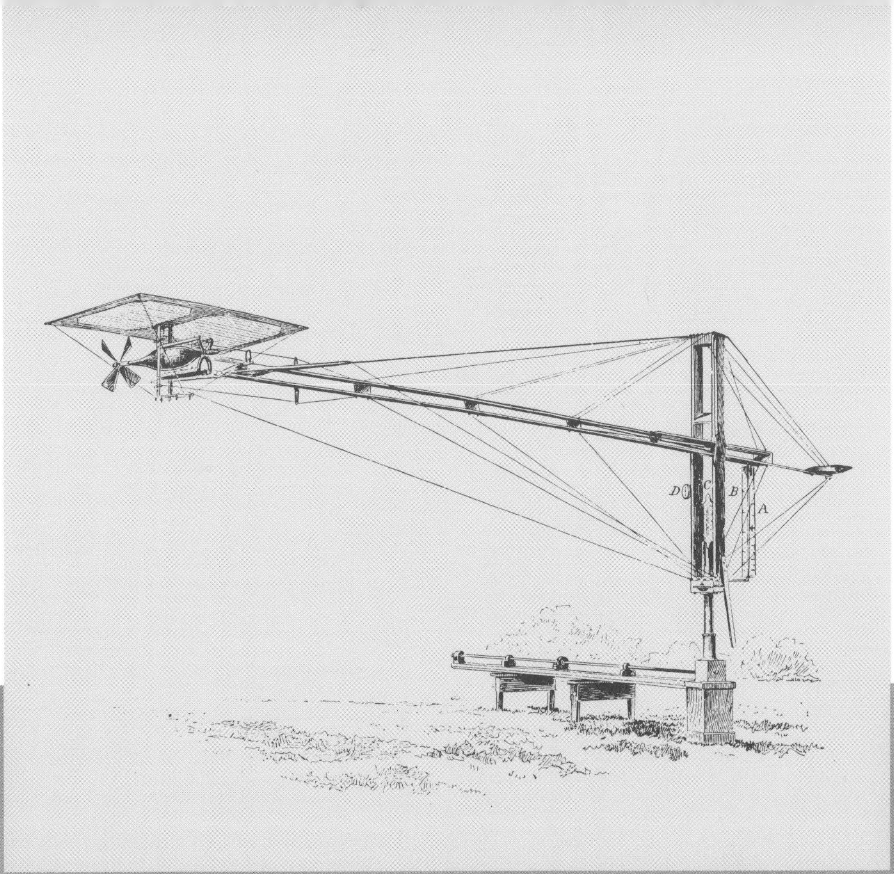
D
C
B
A

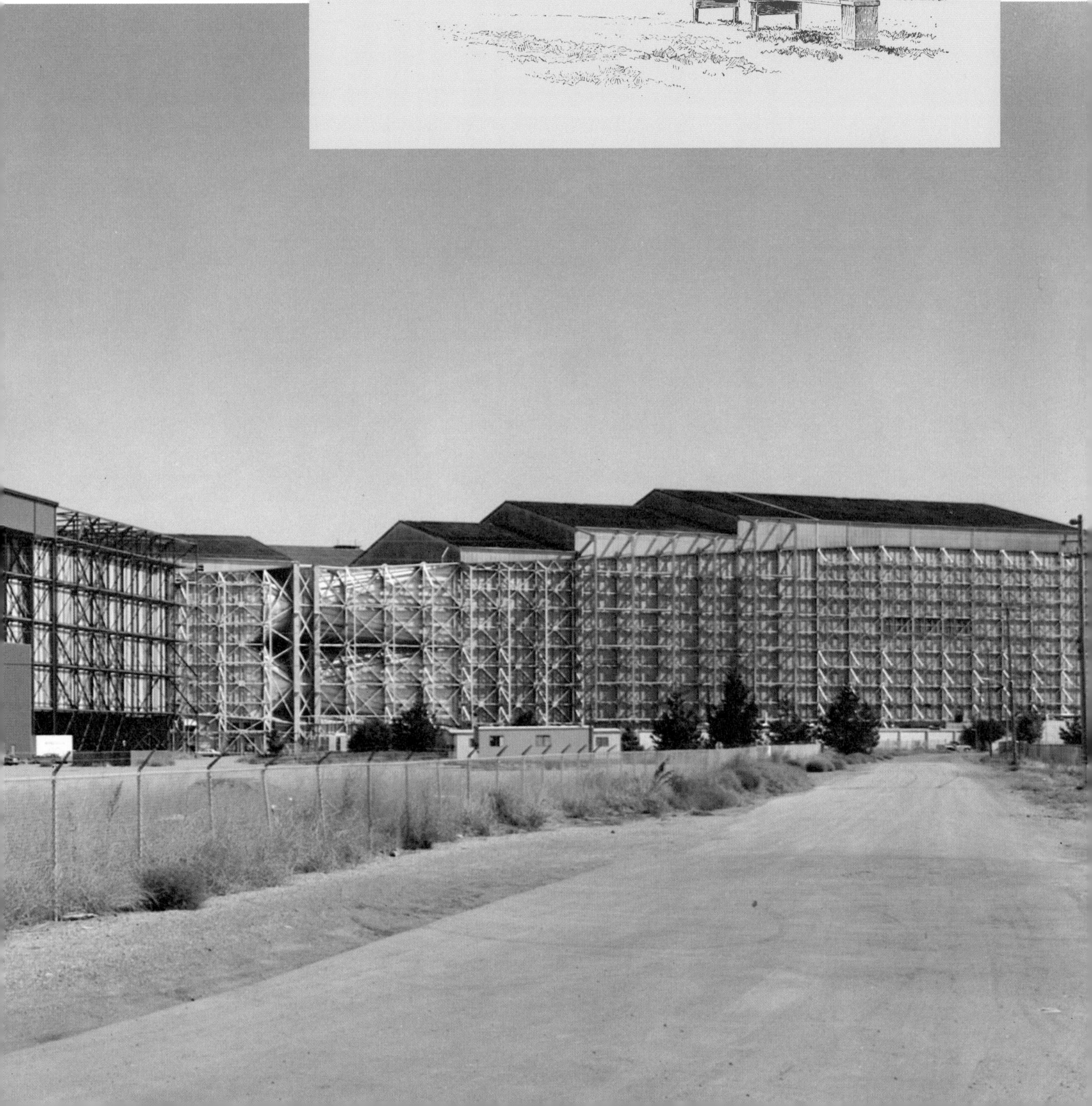

entering the atmosphere at 17,000 miles per hour and gliding
to an unpowered landing at 250 miles per hour. To duplicate
those conditions inside a tunnel requires all the ingenuity one
can muster, and still there is much about the aerodynamics of
the shuttle that tunnel testing does not reveal.

Enter the computer, or rather, a special kind of computer.
One cannot solve the Navier-Stokes equations directly, because of their complexity. But one can use the equations to
create a mathematical model of fluid flow, just as the wind
tunnel creates a physical model. A designer lays a grid over a
proposed design and for a given set of conditions calculates
the forces at each point on that grid, using a simplified form
of the Navier-Stokes equations (figure 7.8). In this technique
the designer in effect samples sets of air particles at discrete
points around the object and calculates the forces affecting
those particles at that spot over a discrete span of time. The
calculations are based on subsets of the Navier-Stokes equations for values restricted to those specific points. The result
is the same kind of information obtained from a wind tunnel
but not plagued by scale factors and accurate over any range
of speeds, pressures, and temperatures.

One needs a very powerful computer to make this technique
work. Carrying out this technique requires a mesh of hundreds of thousands of modeled particles and means evaluating
a complex mathematical expression at each point on the grid.
To simulate the forces over time, this process must be repeated over and over. Thus, analyzing a typical wing at a given
velocity might require several billion arithmetic operations. To
be of practical use to a designer, a computer has to work fast
enough to produce these results in a reasonable amount of
time, somewhere between a few hours and a week of grinding
out numbers.

The idea of computational fluid dynamics occurred to many
persons almost from the day the first electronic computers
began working in the late 1940s. But the processing-speed
requirements always seemed just over the horizon of the
technology available to them, and wind tunnels kept getting
better and better. Throughout the 1950s and 1960s aerody-

7.8
Computational fluid dynamics begins with constructing a mathematical model of the craft to be
tested. A grid is then superimposed on the region
around the craft. The computer solves the equations of fluid flow at each point where the grid
lines intersect. (Photo: NASA)

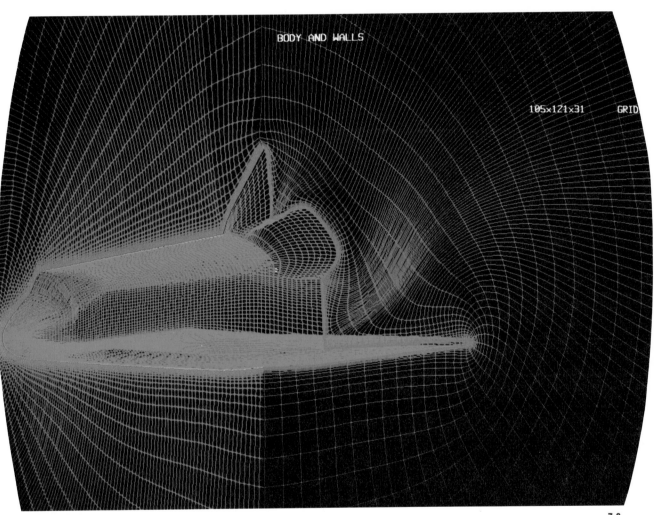

7.8

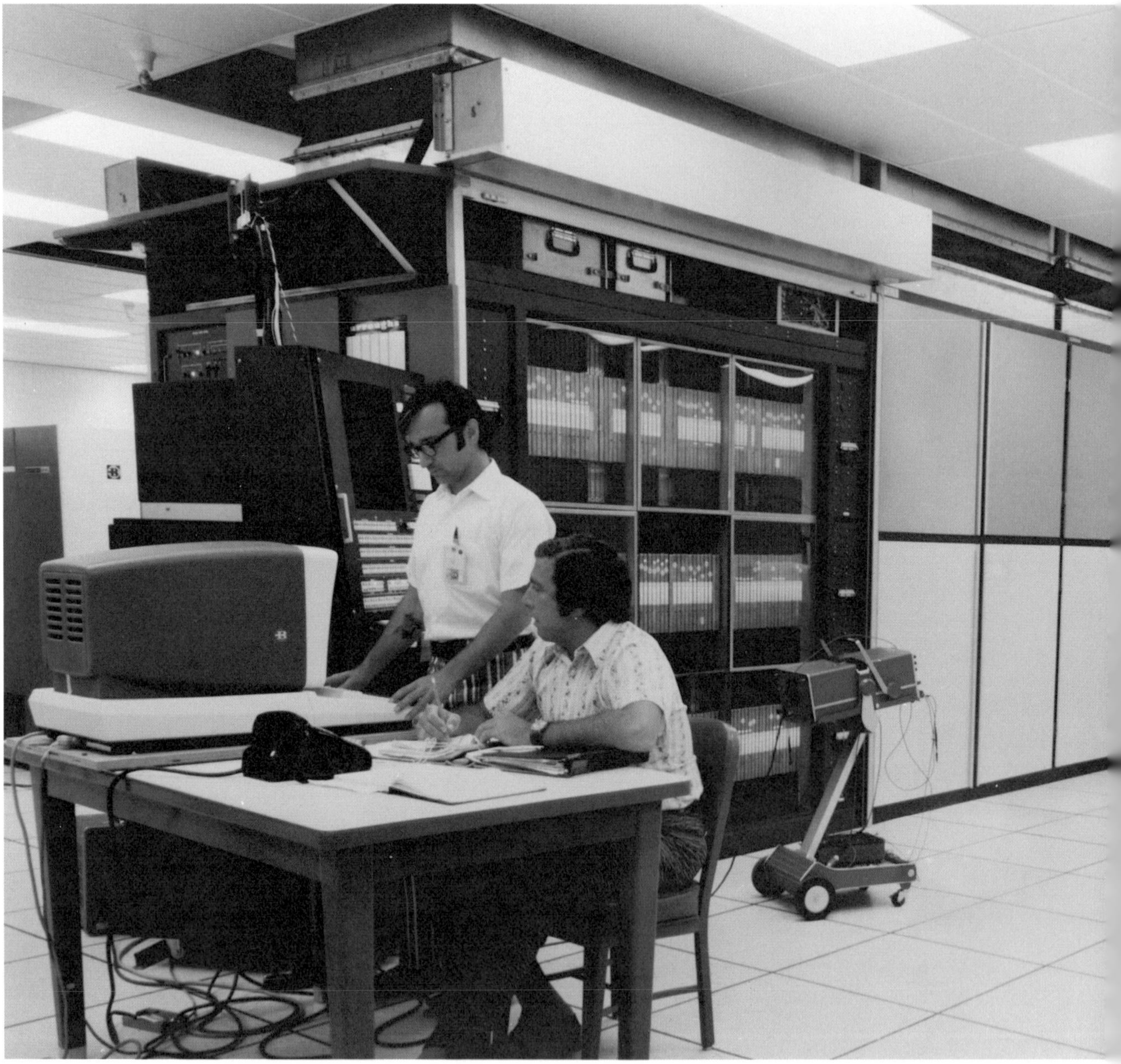

7.9
This early supercomputer was designed at the University of Illinois, built by the Burroughs Corporation in Paoli, Pennsylvania, and installed at the NASA Ames Research Center in Mountain View, California, in 1972. It has sixty-four processors linked to one another and pioneered in what is now known as parallel processing. (Photo: NASA)

namicists made preliminary forays into computational fluid dynamics, keeping an eye on the steady increase in computer speeds. Meanwhile, the complexity of new aircraft designs placed a burden on the wind tunnels: whereas it took about one hundred hours of wind-tunnel time to test the DC-3 design in the 1930s, the Shuttle required over a hundred thousand hours of wind-tunnel testing. Despite advances in wind-tunnels, aerodynamicists were using them to the limits of their capabilities.[21]

The NASA Ames Research Center in Mountain View, California, attempted to accelerate this process in the late 1960s by contracting for a special computer having a design uniquely suited to fluid dynamics problems. Designed at the University of Illinois and built by the Burroughs Corporation, Illiac-IV (the fourth digital computer designed at the University of Illinois in Urbana) used not one but sixty-four separate processors to carry out calculations on grid points simultaneously (figure 7.9). Illiac was hard to program, but it convinced NASA that the concept of computational fluid dynamics was a valid alternative to the wind tunnel. As Illiac and other supercomputers became available, computational fluid dynamics gained advocates.

In 1976 a small company located in Chippewa Falls, Wisconsin, announced a computer that they claimed was the fastest in the world. The company was Cray Research, named after its founder and chief designer, Seymour Cray. The computer was called Cray-1, and it was, indeed, fast: capable of over a hundred million arithmetic operations per second. That meant it could solve in a few hours what other computers required weeks or months to do. This product, and its successor Cray X-MP, finally made computational fluid dynamics a practical complement to wind tunnels for aerodynamics research and design.

Prior to forming the company that bears his name, Seymour Cray worked for Sperry Univac and then Control Data Corporation, and during those years he doggedly pursued the ultimate in computer performance. The Cray-1 was typical of his

7.10

The Cray-1 was not the first supercomputer, but it was the first practical one. (Photo: Cray Research)

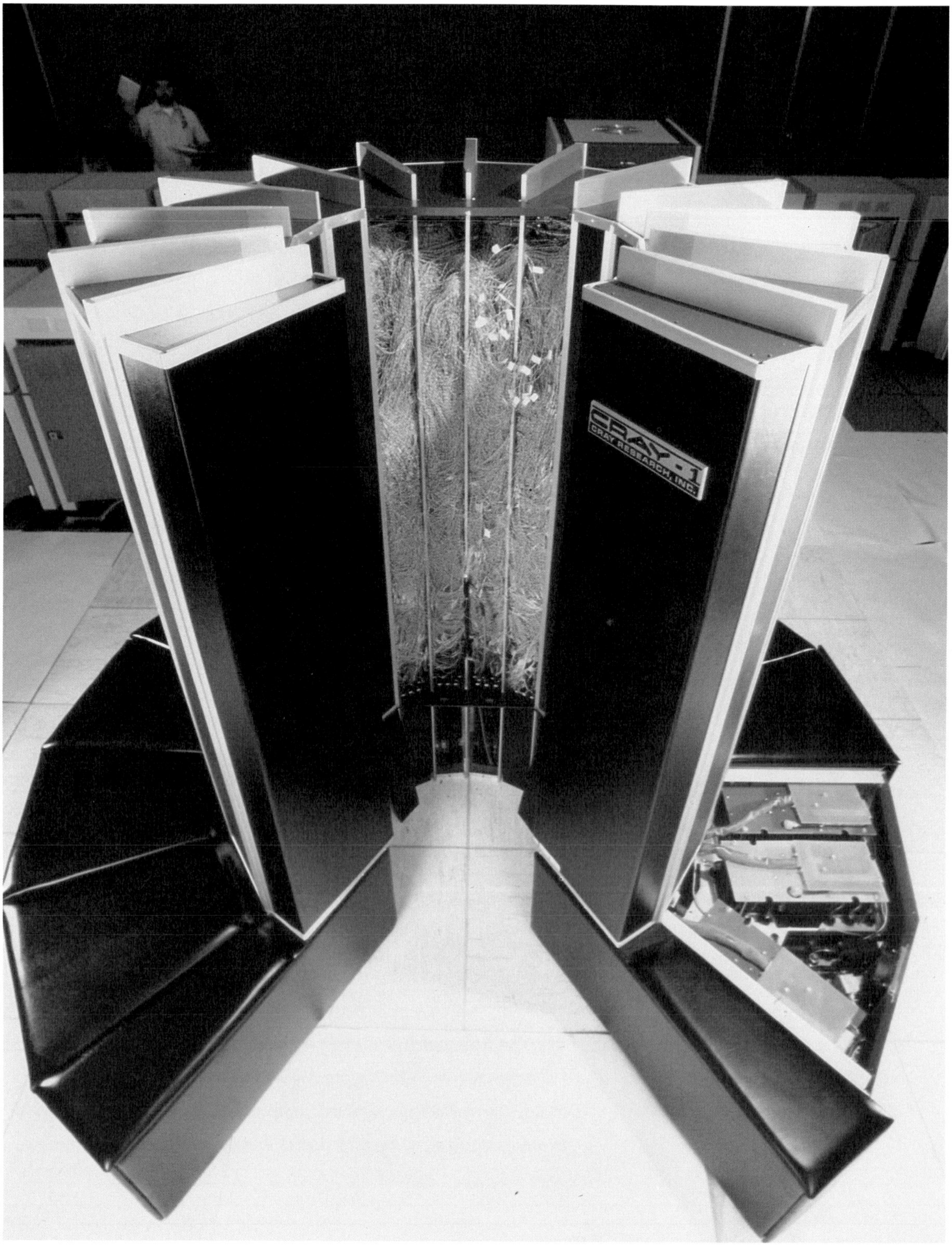
CRAY-1
CRAY RESEARCH, INC.

approach to computer engineering: everything about it was geared toward processing speed. It was a complicated machine, but at the same time no more complicated than necessary (the same elusive property is a goal for aircraft designers as well). The distinctive shape of the computer is a good illustration of Seymour Cray's design philosophy. The processing unit is a tall cylinder open at one side and on the top (figure 7.10). This arrangement ensures that the distance, and therefore the time, a signal spends in transit from any one circuit to any other is a minimum and still allows wirers to install the maze of wiring. No individual piece of wire is more than two feet long, yet there are over sixty-six miles of wire in a complete Cray-1. (A group of workers assemble, wire, and test the wiring by hand in a modest plant in Chippewa Falls.) This arrangement saves only a few billionths of a second over a conventional arrangement of circuits, but in the supercomputer business, a nanosecond (one billionth of a second) is a very long time (figure 7.11).

The basic version of the Cray-1 sold for about $5 million, and the company had little trouble selling all it could manufacture. The first customers were the U.S. and British atomic energy organizations, which used the computers to design nuclear weapons. There the machines computed the propagation of shock waves through nuclear material, a very different application from aerospace but one that uses essentially the same kind of mathematics. Next in line was the intelligence community, which used Crays to decode intercepted radio messages picked up by eavesdropping satellites. After them were the first customers not under a tight veil of secrecy: aerospace companies and NASA.

Cray machines appealed to aerodynamicists because they were reliable and easy enough to program, which allowed aerodynamicists to concentrate on aerodynamics and not on the computer. One had to be conversant with advanced concepts of computation to use it effectively, but Cray-1 provided fast processing speed without the headaches associated with getting it to work. Aerospace designers had enough problems as it was, and they wanted a tool that could help them solve those problems. The Cray machine filled that need.

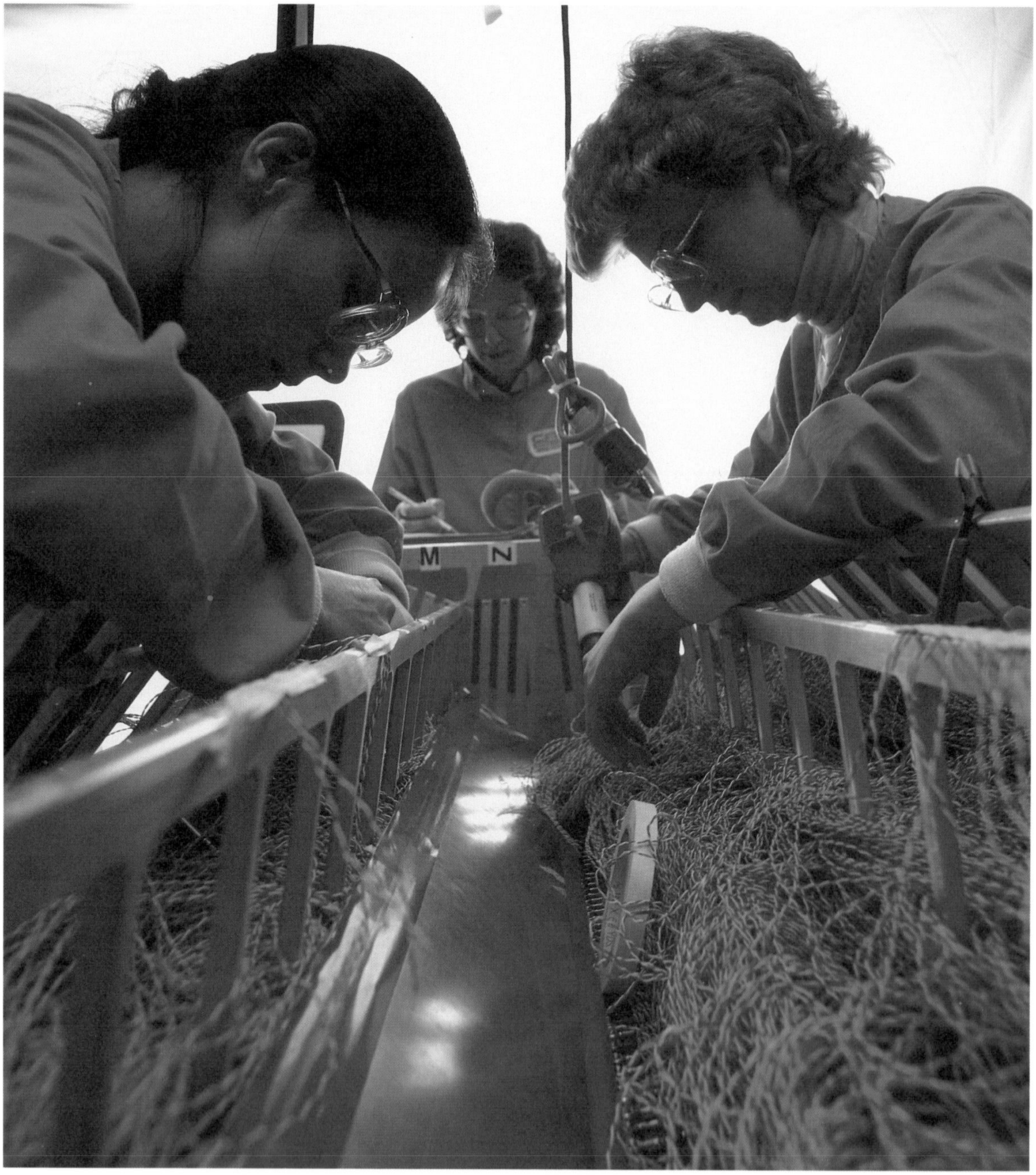
M
N

7.11
At Cray's plant in Chippewa Falls, Wisconsin, workers solder by hand the thousands of short wires that connect the circuits. This method, though very time-consuming and labor-intensive, yields faster processing speeds than printed circuits can give. (Photo: Cray Research)

The Basics of Computational Fluid Dynamics

At the NASA Ames Research Center, supercomputers and wind tunnels exist side by side as equally important aerodynamics research tools. Both types of tools are expensive, and administrators there have gone to great lengths to make these tools available to those who need them and thus make them pay their way. For computational fluid dynamics they do this by dividing up the work so that the supercomputer does only what it can do best—evaluate the expressions derived from the Navier-Stokes equations—while other, less-expensive machines do the rest of the job. At its computer center several supercomputers toil away on fluid-dynamics equations. Once the solution for a given case is computed, it is transmitted to a minicomputer (typically a Digital Equipment Corporation VAX) via a high-speed network. A researcher can then call up this solution on a dedicated terminal that displays the answer in picture form, which conveys much more of a sense of the problem and its solution than a table of numbers would convey. The researcher can examine the solution, manipulate the image, zoom in on a particular area of interest, and so on. Meanwhile, the supercomputer can go on to work on other problems, so it is always kept busy.

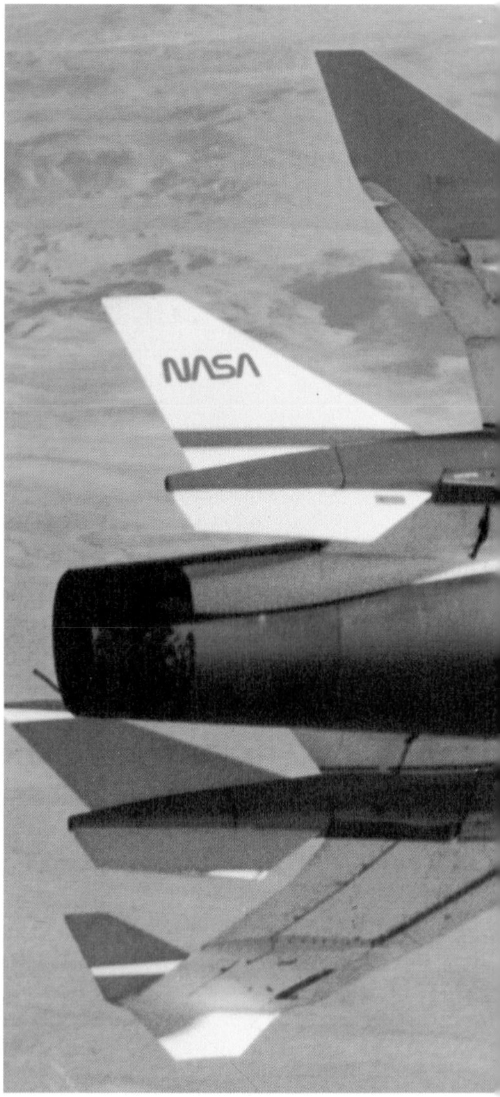
NASA

In the late 1970s this approach began to pay off. One early success was the experimental NASA aircraft called HiMAT (Highly Maneuverable Aircraft Technology), designed to test concepts of high maneuverability for the next generation of fighter planes (figure 7.12). Wind tunnel tests of a preliminary design for HiMAT showed that it would have unacceptable drag at speeds near the speed of sound; if built that way the plane would be unable to provide any useful data. The cost of redesigning it in further wind tunnel tests would have been around $150,000 and would have unacceptably delayed the project. Instead, the wing was redesigned by computer at a cost of $6,000.

Another example of success is retrofitting the Boeing 737 short-range passenger jet with new fanjets (see figure 7.4 above). The fanjets, though quieter and more efficient than the plane's original engines, are wider than the older units.

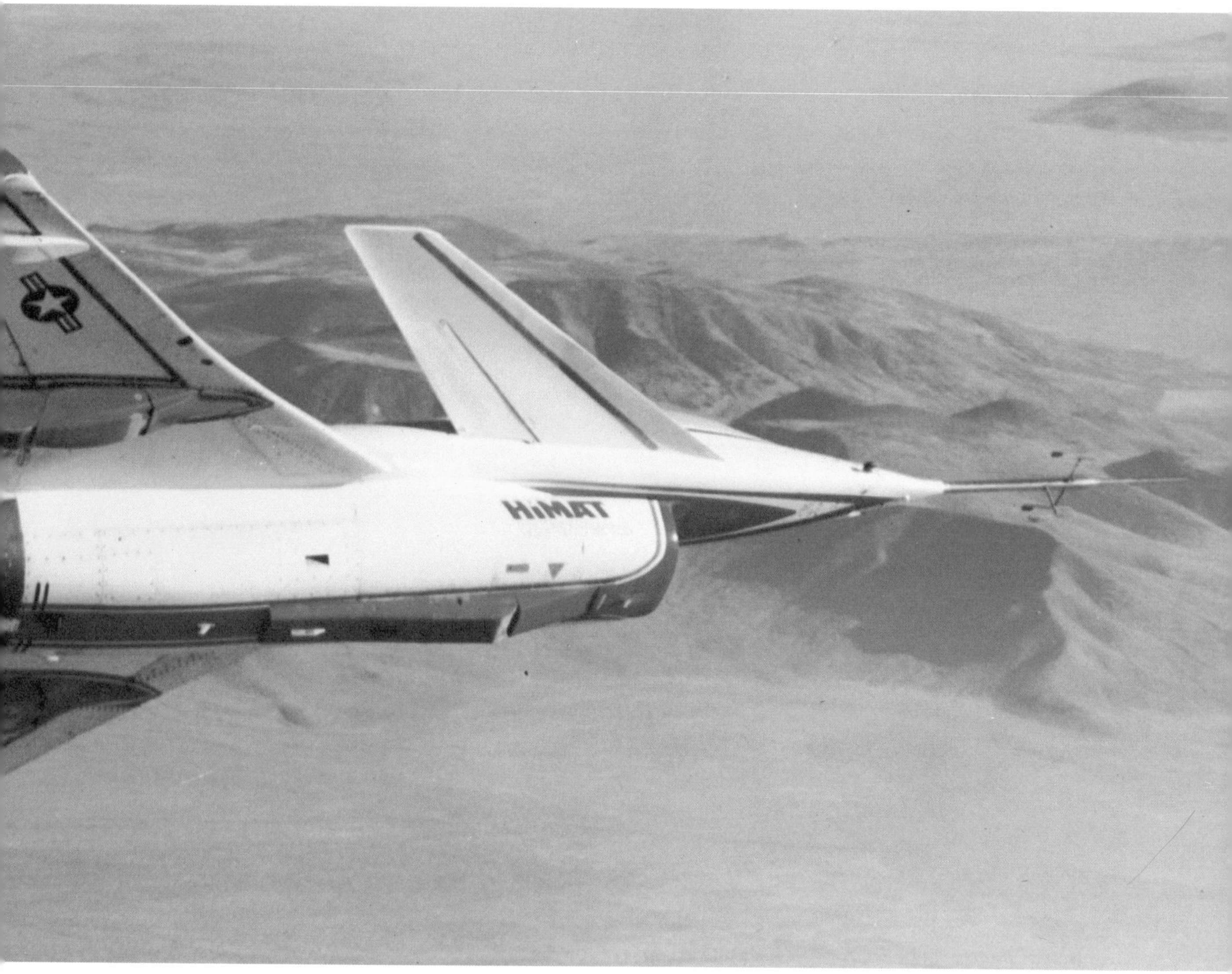
HiMAT

7.12

HiMAT (Highly Maneuverable Aircraft Technology) was a one-third scale model of a fighter intended to test new concepts in aerodynamics and controls. It was flown by a pilot on the ground operating conventional controls and instruments. (Photo: NASA)

That potentially meant greater aerodynamic drag. It also seemed that using fanjets would require a longer and more complicated landing gear to prevent the engines from hitting the runway. But Boeing engineers were able to design the engine cowlings in a way that avoided the need for a longer landing gear. With the help of a Cray X-MP they slightly flattened the air intakes and thereby got the engines to clear the runway while keeping drag to a minimum.

Current activity in computational fluid dynamics includes the planned Aerospace Plane, which will take off from an ordinary runway and ascend directly to a low earth orbit, and a redesign of the combustion chamber of the Space Shuttle main engine, to name only two (figure 7.13). Computational fluid dynamics is by no means restricted to the air. Water too is a fluid, and its effect on ships' hulls can be analyzed the same way. The hull of *Stars and Stripes,* which in 1987 returned the America's Cup to the United States, was designed with the help of supercomputers at Grumman Aerospace Corporation and at a Cray Research laboratory in Minnesota. It is already evident (to the dismay of many sailors) that the next winner of the America's Cup will be the one with the better computer as well as the better boat.

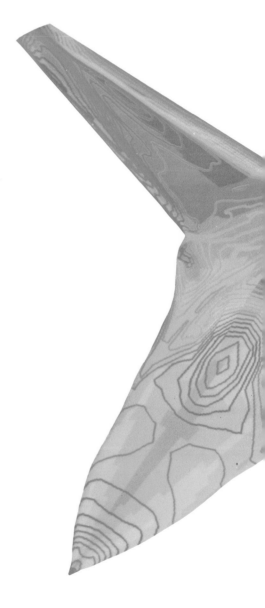

Conclusion

The recent history of the design of air and space craft reveals an accelerating chase of greater performance demands followed by advances in design and manufacturing techniques. The computer has driven most of the advances in design. Computational fluid dynamics, coupled with advanced computer graphics software, has been one of the biggest successes. Putting designs into volume production at a reasonable cost has been a more elusive goal. To translate an idea into hard metal, one can turn to powerful computers applying advanced mathematical techniques, yet despite all that, there is still a need for the contriving brain, the skillful hand, and the perceptive eye.[22]

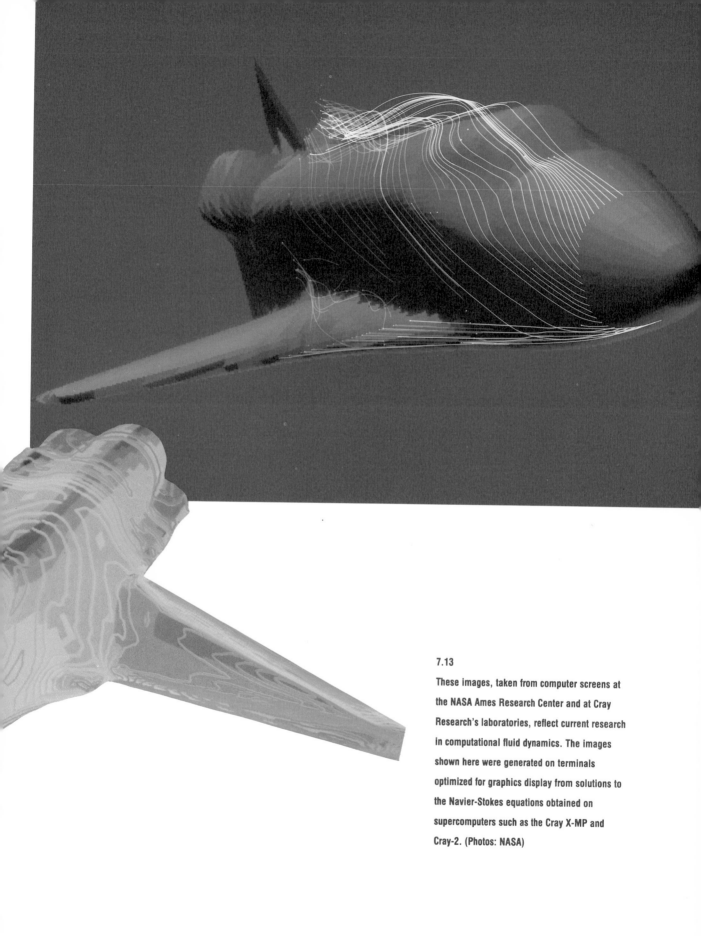

7.13

These images, taken from computer screens at the NASA Ames Research Center and at Cray Research's laboratories, reflect current research in computational fluid dynamics. The images shown here were generated on terminals optimized for graphics display from solutions to the Navier-Stokes equations obtained on supercomputers such as the Cray X-MP and Cray-2. (Photos: NASA)

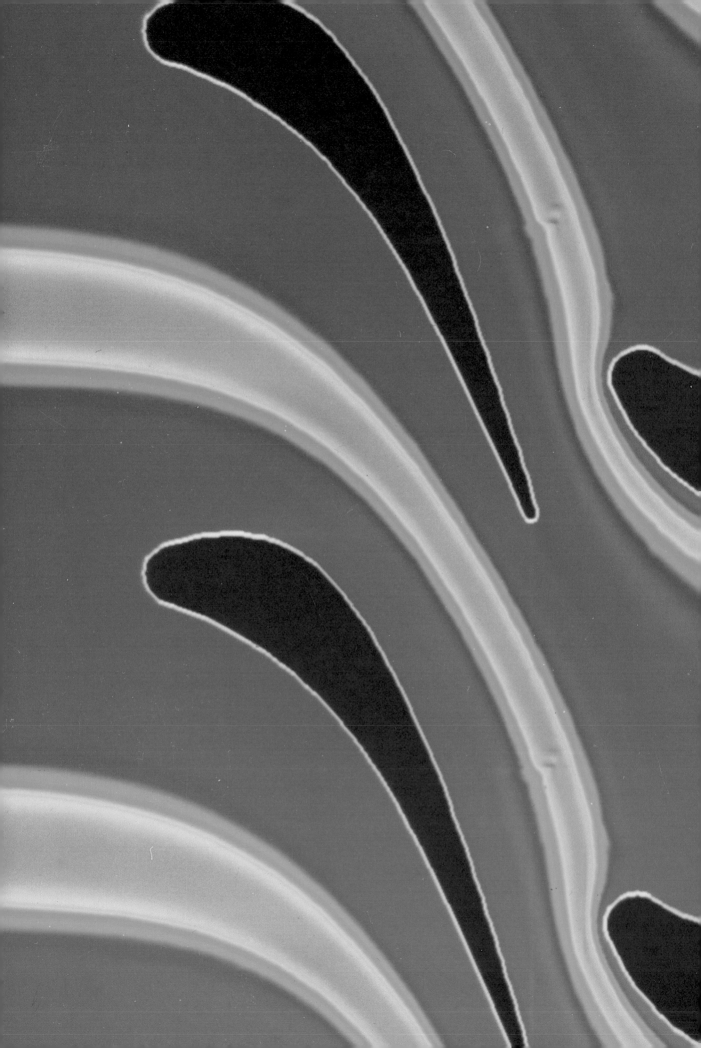

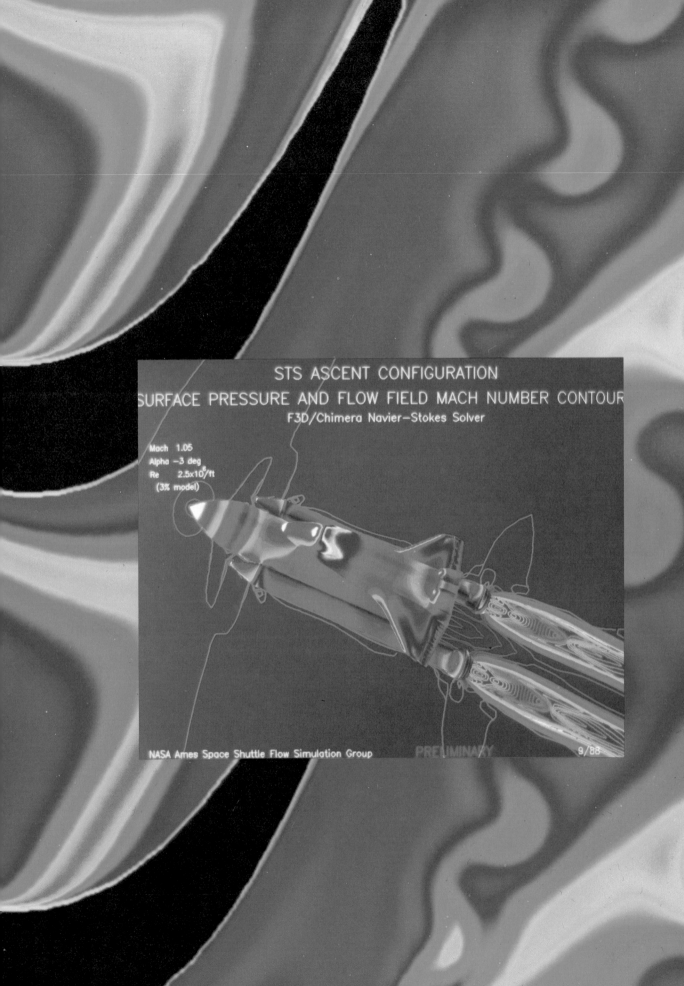

STS ASCENT CONFIGURATION
SURFACE PRESSURE AND FLOW FIELD MACH NUMBER CONTOUR
F3D/Chimera Navier–Stokes Solver

Mach   1.05
Alpha −3 deg
Re 2.5×10^6/ft
(3% model)

NASA Ames Space Shuttle Flow Simulation Group   PRELIMINARY   9/88

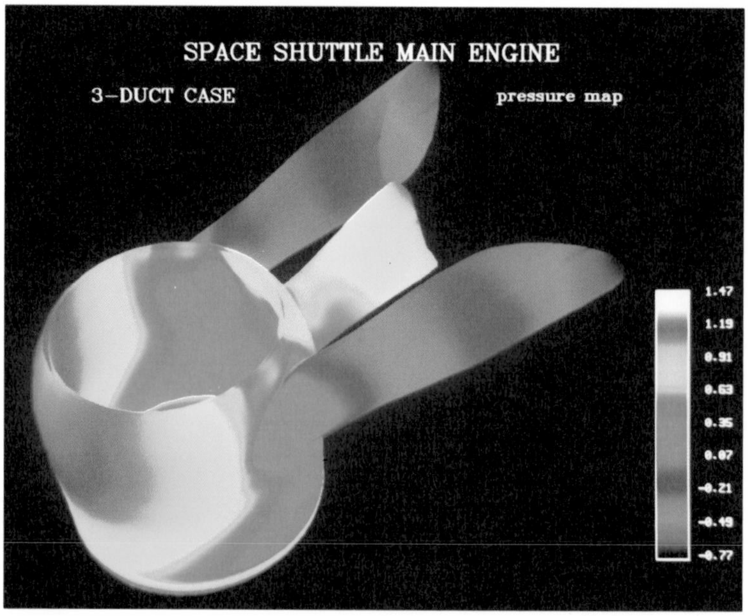
SPACE SHUTTLE MAIN ENGINE
3-DUCT CASE
pressure map
1.47
1.19
0.91
0.63
0.35
0.07
-0.21
-0.49
-0.77

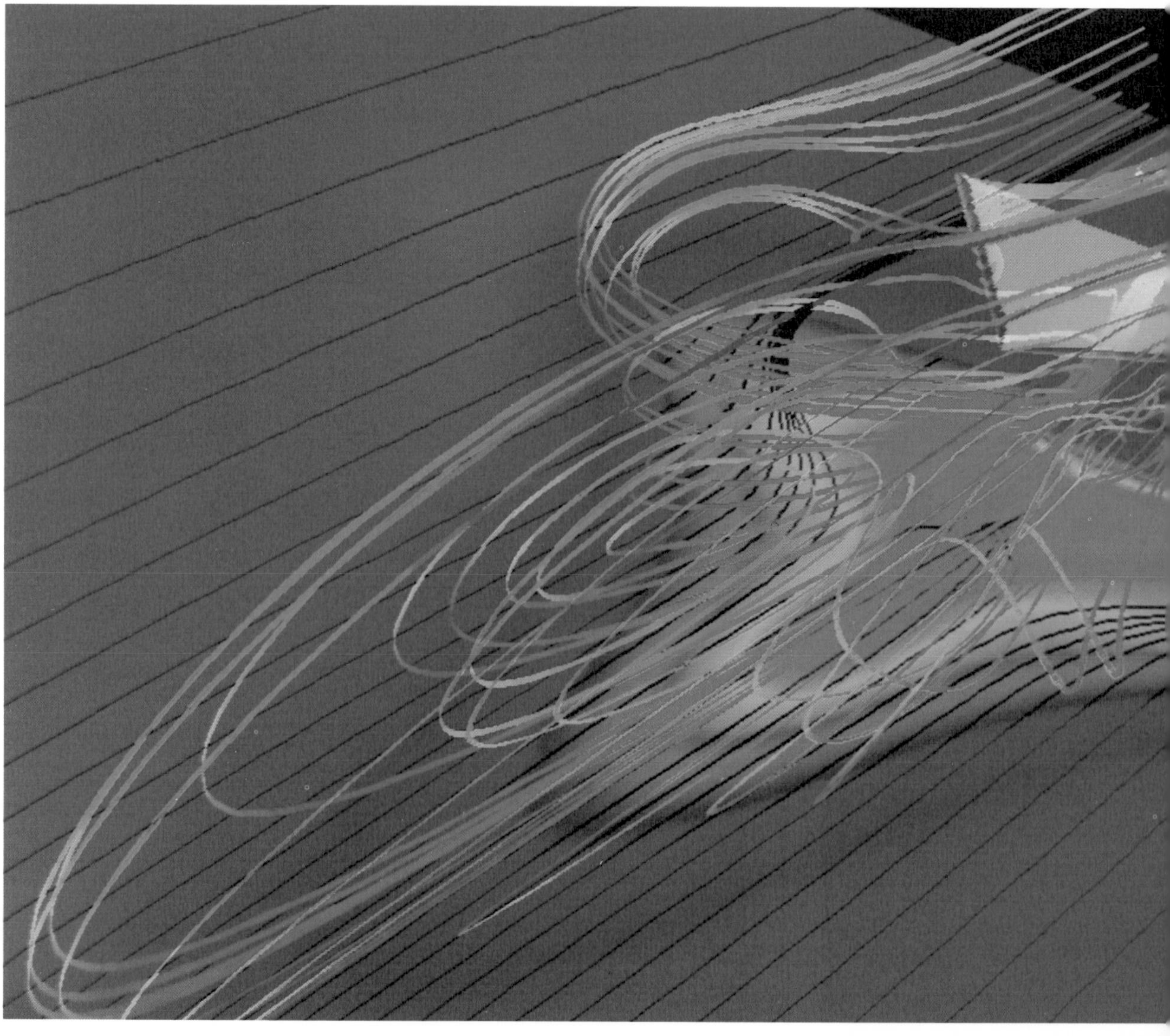

INSTANTANEOUS PRESSURE DISTRIBUTION
IN A TURBINE STAGE

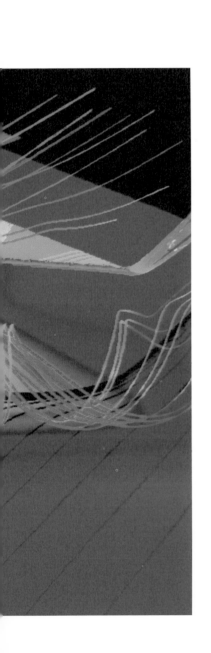

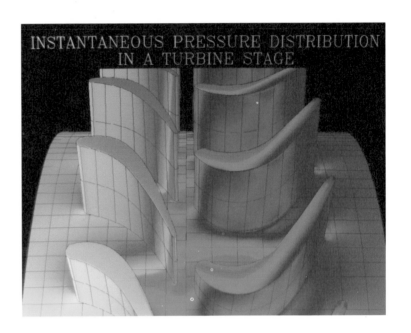

Chapter 8
Advances in Simulation, Testing, and Control

Once a flying machine is built, the computer takes on a different set of tasks. These new duties call on its ability to do real-time calculations and to display results in graphic form, products of the past twenty years of computer science and engineering.

Even with all the techniques of CAD/CAM, wind-tunnel testing, and computer analysis of the structure and aerodynamics of a craft, design is still fraught with uncertainty. It does not cease when the first prototype is rolled out of the plant. The final stage of design is flight testing. Here a test pilot checks predicted performance against actual flight data. If all goes well, there will be basic agreement. But if flight tests reveal a huge discrepancy, it is "back to the drawing board," with all the delays and expense that implies. The advanced techniques described in the previous chapter make this less likely to occur; nevertheless, it does happen. In any case, flight testing usually results in some refinement of the initial design.

Closely related to flight testing is training pilots and astronauts to fly the new designs (for unmanned spacecraft there is a similar need to train ground controllers to run the mission through its various phases correctly). Training uses the computer's ability to simulate actual flight, the activity first envisioned for Project Whirlwind. Simulation takes a design and animates it by looking at the machine's behavior through time. In flight testing, the predicted behavior is matched against the actual behavior of a real craft. A simulator trains future pilots and astronauts by mimicking that behavior on a machine on the ground. The machine simulates flight by containing in its programs the designer's calculations, the results of wind-tunnel tests, and data from actual flight tests of the craft if available. At the heart of a modern flight simulator is a fast digital computer, one that can compute in real time and display its results in a variety of forms.

The final place that the computer revolution has had an impact is in the cockpit. Controlling an airplane or spacecraft has never been easy. The digital computer, on-board and linked to flight controls, has cast this old problem in a new light.

Flight Testing: A Science of Measurement

Flight testing is a science of measurement. It requires collecting, recording, storing, processing, and finally analyzing large amounts of data. For years people did these things manually with the aid of a few simple instruments and calculating machines. The advent of high-speed digital computers has not altered the basic purpose or need for flight testing, but it has altered the way in which tests are carried out.

Test pilots are among the most romanticized of all people in air and space flight. True, they do perform the dangerous but necessary task of removing as much uncertainty as possible from a design. Tony LeVier, for many years chief test pilot for Lockheed, put it this way: "There are still some things you can't test for in a wind tunnel, and no airplane can be instrumented to the point of perfection. Up to 90 per cent of any new plane can be proved out before it flies, if you want to spend the money to do it, but it is the test pilot who has to find out the remaining 10 per cent."[1]

In graphic terms, it is the job of the test pilot to probe the corners of a craft's "flight envelope" that is, its *designed* maximum airspeed, altitude, rate of climb, and so forth. When data from a test are plotted on a piece of paper, they fill a region that one might say "encloses" the craft's performance limits—hence the name (figure 8.1). Inside this region there is no problem, but when one gets near, say, the maximum design altitude, the designer's predictions are a lot less trustworthy. Perhaps just before attaining this altitude, the plane is prone to stall or otherwise lose control; perhaps not. The test pilot takes the airplane to the edge of the envelope and finds out.

In Tony LeVier's day the pilot carried a knee pad on which he recorded instrument readings during such a test (figure 8.2). A sixteen millimeter camera mounted in the cockpit or in the nose recorded other instrument readings (figure 8.3). If there was room, on-board equipment racks recorded data on rolls of paper.

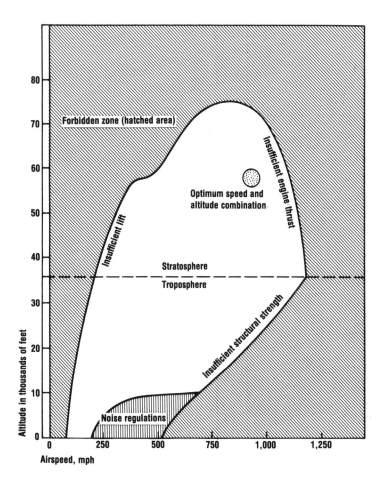

Forbidden zone (hatched area)

Insufficient lift

Insufficient engine thrust

Optimum speed and altitude combination

Stratosphere
Troposphere

Insufficient structural strength

Noise regulations

Altitude in thousands of feet

Airspeed, mph

8.1

When the performance data of an airplane are plotted on a graph, they define a region within which it must stay. Venturing outside that envelope may be either unsafe (because of the inherent design of the craft) or illegal (because of noise or speed regulations).

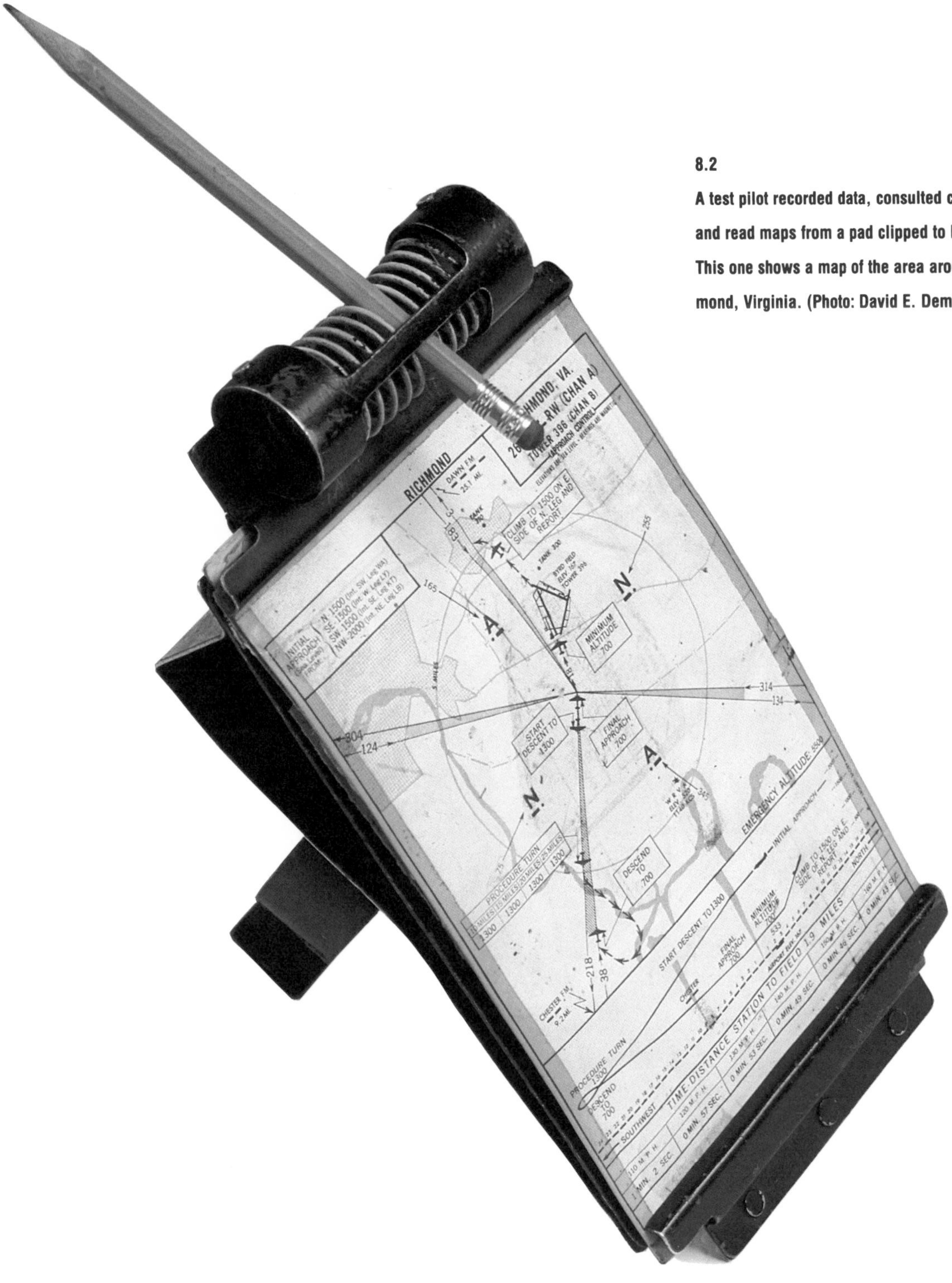
RICHMOND
TOWER 396 (CHAN B)
APPROACH CONTROL
INITIAL APPROACH
MINIMUM ALTITUDE 700
START DESCENT TO 1300
FINAL APPROACH 700
DESCEND TO 700
PROCEDURE TURN
CLIMB TO 1500 ON E. SIDE OF N. LEG AND REPORT
TIME-DISTANCE STATION TO FIELD 1.9 MILES
SOUTHWEST
NORTH

8.2

A test pilot recorded data, consulted checklists, and read maps from a pad clipped to his knee. This one shows a map of the area around Richmond, Virginia. (Photo: David E. Dempster)

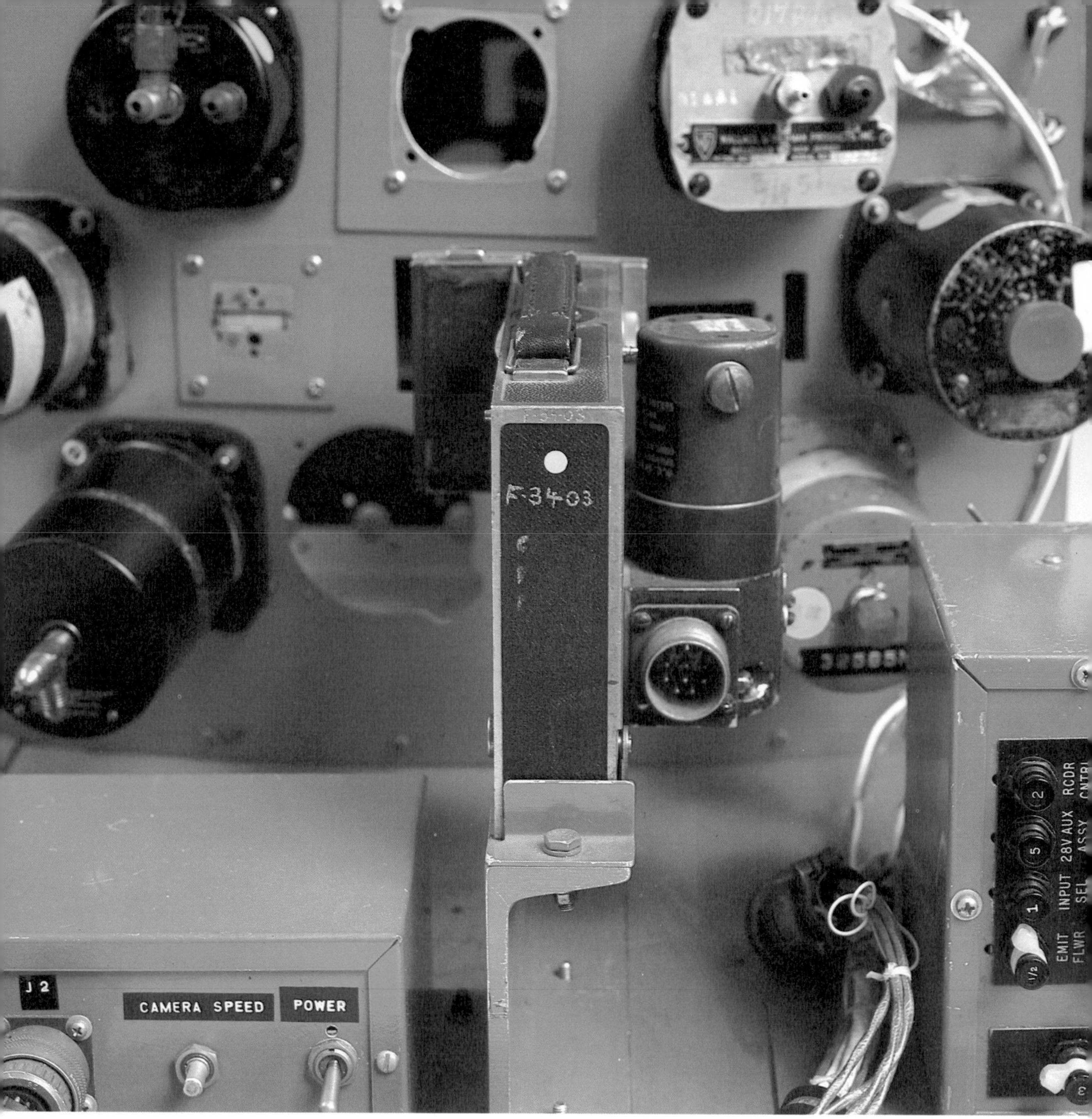
F-3403
J 2
CAMERA SPEED
POWER
EMIT FLWR
INPUT SEL

8.3

A photo panel was a common means of recording test-flight data in the 1950s. The panel shown here was mounted in the nose of a Grumman jet and consists of standard flight instruments, a clock, and a 16-mm camera that recorded the readings of those instruments. After a test flight the film was developed and the instrument readings were plotted, reduced, and analyzed by the pilot and test engineers back on the ground. (Photo: David E. Dempster)

After it was recorded, someone then had to reduce this data and turn it into meaningful information about the plane's performance in that region. Teams of human computers often did this job, but it might just as well have been done by the test pilot himself. Michael Collins, one of the Apollo astronauts, recalled his earlier career as a U.S. Air Force test pilot (figure 8.4): "After each flight, the developed film and oscillograph paper were delivered to us; and weekends and nights would find us hunched over a desk calculator or peering at a film projector, trying to reduce this overwhelming amount of information into a terse report . . . , so that we could move on to the next test series and begin the miserable process all over again. Ah, the glamor of a test pilot's life!"[2]

Collins makes clear how flight testing was an iterative process requiring many, many flights. For this reason more than any other the introduction of computers into the loop had a great effect. Automation of the individual steps of instrumentation, recording, and reduction of data might be beneficial, but it would not change the basic structure of the activity. But as

8.4

Michael Collins (left, shown here with NASA Research Pilot John Manke) was a test pilot who flew this F-104 before joining the Astronaut Corps. He later piloted the Command Module of the first manned lunar landing in July 1969. (Photo: National Air and Space Museum)

8.4

soon as computers with a real-time computing capability appeared (around 1970), the situation changed. Then one could record the data, transmit that data from the plane to a ground station, reduce the data, and make a preliminary analysis *while the airplane was still in the air*. If for some reason the pilot did not get close enough to the envelope edge in question to get meaningful test results, the test engineer on the ground would immediately know and tell the pilot to repeat the maneuver. In Collins's day the pilot would find out a few days later that he would have to go back up and fly the maneuver over again. Real-time computing and telemetry meant that in a single flight the pilot could probe not just one but several corners of the envelope and know before landing that he had gathered the information needed by the flight-test engineers on the ground.

The experimental NASA X-29 aircraft illustrates how all this works in a modern test program (figure 8.5). The X-29 continues the tradition of X-series research planes of proving in actual flight some new design concept. This tradition began with the legendary X-1, which Charles Yeager flew through the sound barrier in 1947. The X-1 explored flight around and just above the speed of sound, something that could not be tested either with existing airplanes of the day or with wind-tunnel tests.

The X-29 continues this tradition. The plane was designed mainly to test the concept of sweeping the wings forward instead of back, as in most fighter jets. Aerodynamicists have known for decades that sweeping the wings forward can have the advantages of ordinary swept wings at high speeds, while also making the craft easier to maneuver and control at lower speeds. But the problem of making the wings light and stiff enough to resist bending prevented the idea from being adopted. The X-29 explores the question of whether advanced composite materials (graphite fiber and resin) can make this concept practical.

But unlike previous X planes, the X-29 is also a test-bed for several other innovations. Doing that saves time and money, and yields valuable data about the interaction of one design innovation with another. Yet it violates the long-standing rule of flight testing that you test only one thing at a time to prevent uncertainties in one area from interfering with gathering reliable information about another area. The X-29 deliberately violates this rule, but with no ill effects, thanks to the extensive use of computers in its test program.

Besides having a forward-swept wing, the X-29 incorporates fully digital active controls (described later), a forward-mounted horizontal stabilizer (called a canard), a thin wing with a "supercritical" shape, and a design called "aeroelastic tailoring," by which the wing's shape adapts itself to given flight conditions and thus changes its aerodynamic properties in precisely specified ways. None of these concepts are really new, but testing them as a coordinated system in a single aircraft is a new idea. Since its first flight in the fall 1984 the X-29 has been testing each of these innovations individually and also in interaction with one another in a single craft.[3]

The Air Force is interested in the X-29 because this plane, like the HiMAT, explores ways of making aircraft more maneuverable at both low and high speeds. In aerial combat a fighter must be agile and fast, two qualities that often conflict with one another. Large and heavy engines make a plane fast, but the weight of those engines limits agility. Controls designed for very high speeds may not work at lower speeds. The forward-swept wings and other features of the X-29 explore ways of making a plane agile in the transsonic region (around the speed of sound), where military strategists believe most air-to-air combat will take place. The hope is that the forward-swept wing will reduce drag around the speed of sound just as a rear-swept wing does, but in addition, that it will retain its lift at lower speeds and at higher angles of attack. The canards, the digital controls, and the supercritical wing each contribute to greater agility in other ways.

Whether this design will prove out remains to be seen; that is the purpose of the program. Grumman Aerospace Corporation of Bethpage, New York, built two identical X-29 airplanes.

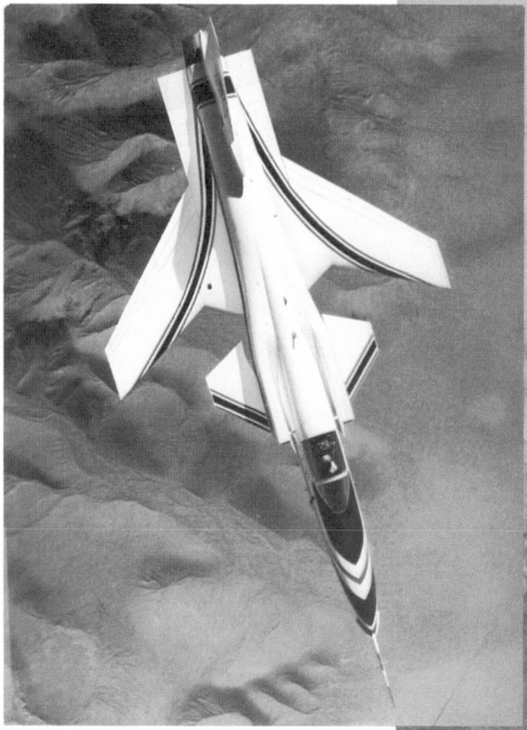

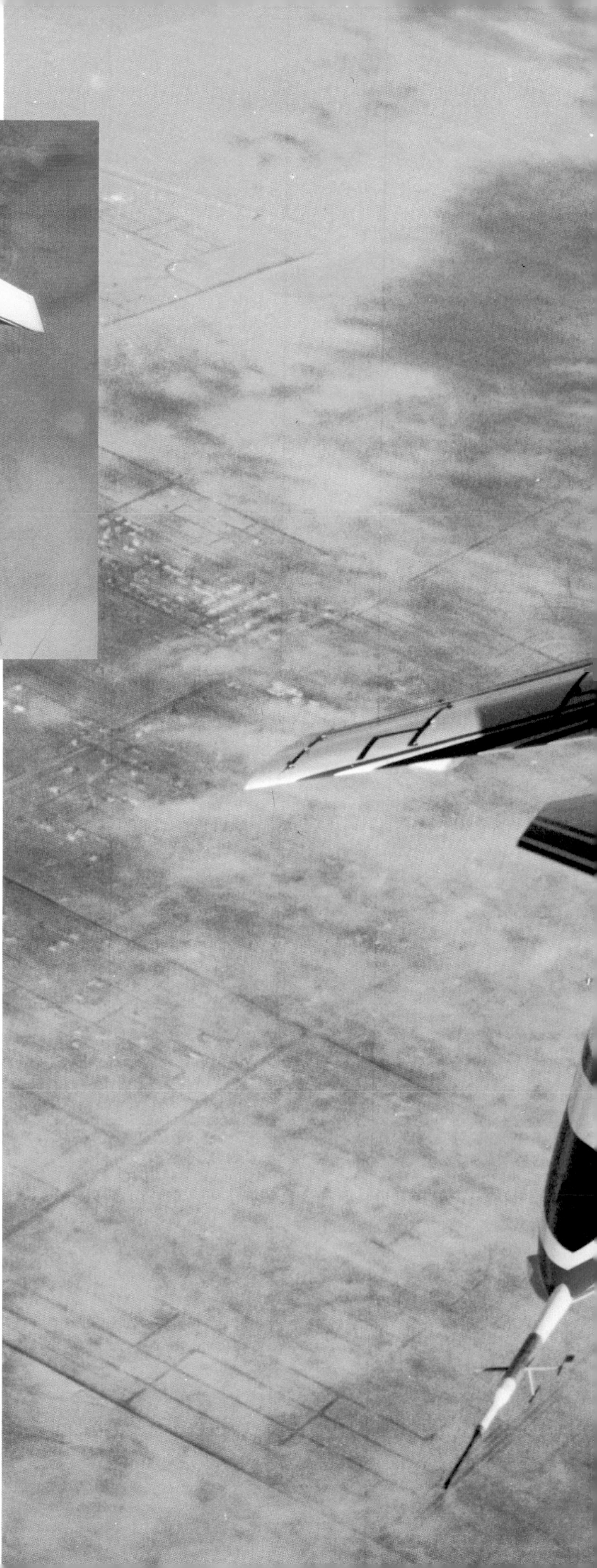

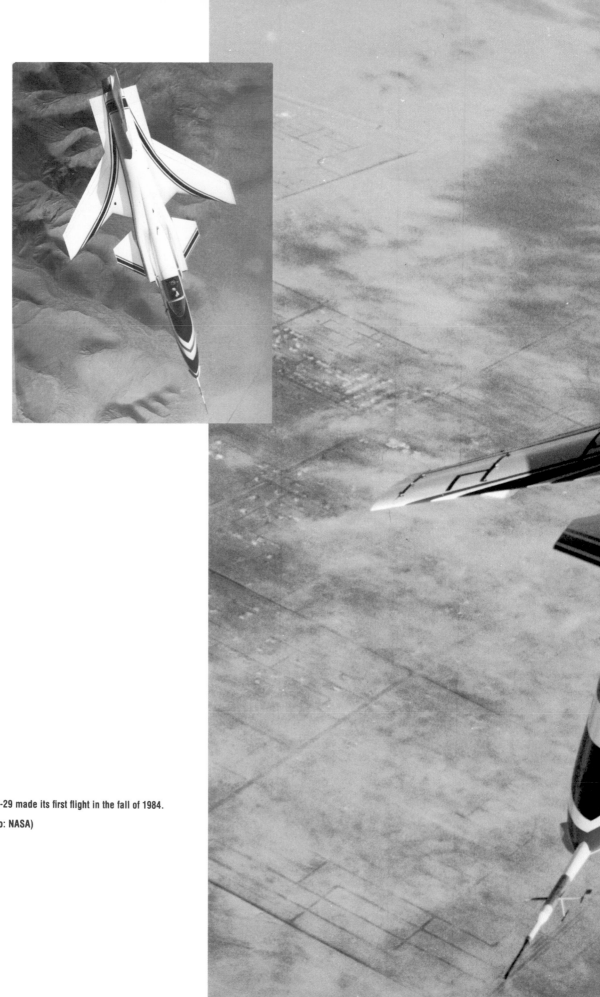

8.5

The X-29 made its first flight in the fall of 1984. (Photo: NASA)

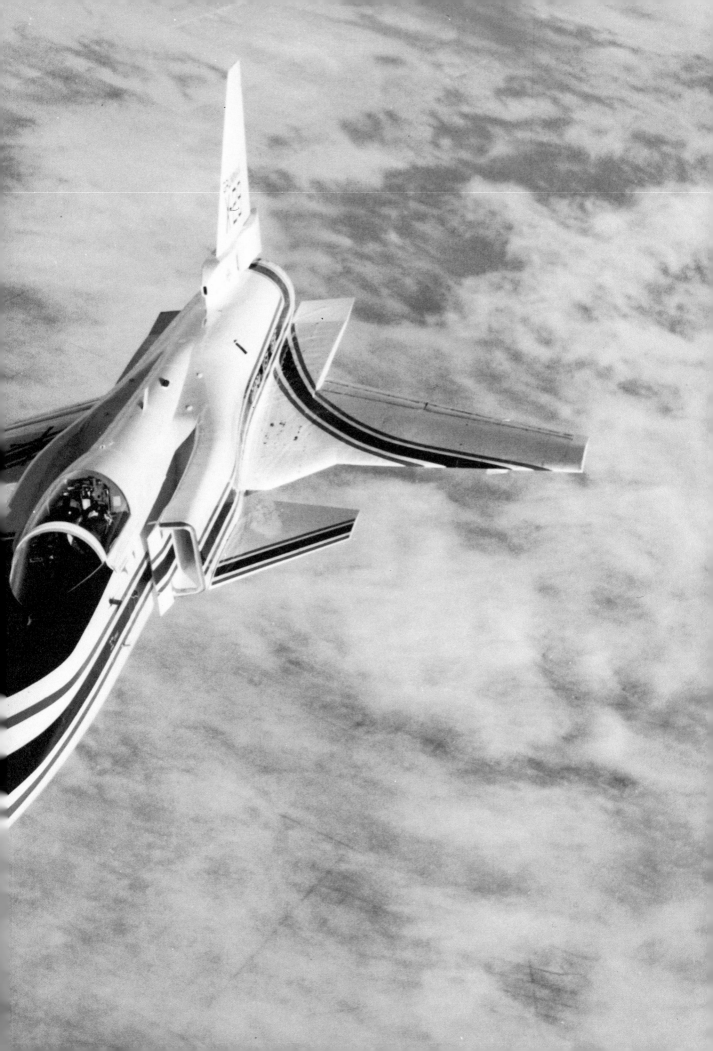

Grumman X-29 Data Path
NASA
Dryden, CA
Grumman
Calverton, NY
Bit Rate
Up
Converter
RCA
Earth
Station
PCM
NASA
Dryden
Ground
Station
Grumman
Interactive
Processing
RCA
Earth
Station
Bit Rate
Down
Converter
VAX
11/750
Failure
Status
Monitor
TP
CYBER
172
R/T Analysis
Station
Interactive
Analysis
Stations

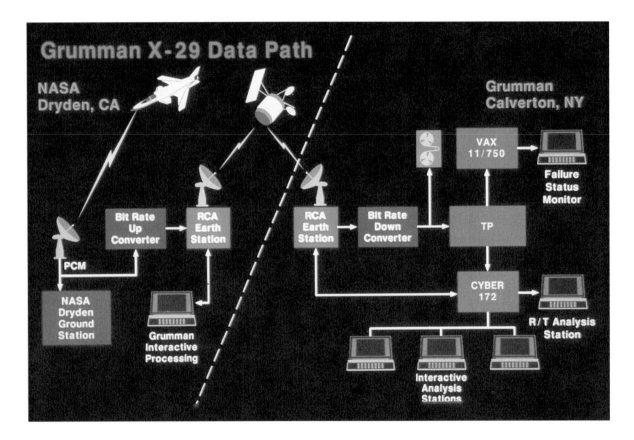

8.6

Testing the X-29 is done in real time. Preliminary data are reduced and analyzed while the plane is in flight. Data reduction and analysis are carried out not only at Edwards Air Force Base in California, but also at Grumman's facility in Calverton, New York, where the plane was designed and built. Telemetry and satellite communications make this whole process transparent to the test engineers at Grumman. (Grumman Aerospace Corporation)

NASA jointly sponsors the program with the Defense Advanced Research Projects Agency and flies them at its Dryden Research Center located adjacent to the famed Edwards Air Force Base in California. Test flights are carried out over a dry-lake bed, and data are transmitted to a computer facility at Edwards and also via satellite to Grumman's Long Island facilities (figure 8.6). All testing is done in real-time, with two and sometimes even three flights a day when everything goes well. The test pilots are still very much part of this activity, and they continue to push the limits of performance, even if not quite as the popular press and Hollywood movies have portrayed them as doing. But along with the test pilots, the engineers, and the ground support crew is a new partner, the computer.

The Digital Takeover of Flight Simulators

Chapter 5 described how Project Whirlwind began as a plan to construct a flight simulator for training Navy pilots. Many of the ideas in the original proposal finally took form in the mid-1970s, when advances in digital computers were applied to flight simulators. Here again is a brief description of the context in which flight simulation evolved.

A designer develops a mathematical description of how an airplane will behave in actual flight. Aerodynamic theory, wind-tunnel tests, and computer tests give him the equations of lift and drag; engine performance data give him a description of thrust and top speed; and so forth. During flight testing, these equations are checked against actual performance and further refined if necessary. These equations describe how a craft will behave for a given set of conditions. For example, if a passenger plane has a full load of passengers and fuel and is flying at thirty-five thousand feet and the pilot turns the control yoke a certain amount, the equations will describe the rate at which the plane will turn.

These same equations of the dynamics of flight can be solved on the ground using hypothetical data, and the solution displayed on a computer terminal or other device. Here the equations simulate, rather than describe, the behavior of the craft. Simulation has two important benefits for aerospace: First, it can help predict the performance of a craft in its myriad details well before committing resources to production of a prototype. Second, it can be used to train pilots to fly a new model airplane without the expense and danger inherent in using the real plane for this purpose. For these reasons simulation is widespread throughout aerospace and takes up a major proportion of the time of pilots, engineers, and support crew.

Implicit in the notion of simulation is that one has a machine that can solve the equations with sufficient accuracy and speed to recreate the feel of the real thing. One does not need an electronic computer to do this, but a high-speed digital computer and a method of displaying high-quality graphics and pictures form an ideal tool with which to construct realistic simulators.

The first simulators were little more than mock-ups of cockpits. Levers, wires, and hydraulic valves connected to the stick and rudder pedals translated a trainee's actions on them into changes on the cockpit instruments, as if the machine were actually flying. Edwin A. Link, who learned to fly shortly after Lindberg's transatlantic flight in 1927, built a simulator using pneumatic hoses and mechanisms borrowed from his father's pipe-organ factory in Binghamton, New York. He took it to county fairs, where it attracted quite a bit of attention. (There is still a close relationship between the multimillion dollar simulator business and the video games sold for the home computer or found in the local shopping mall.)

In 1934 the U.S. Army Air Corps purchased several Link trainers for training pilots to fly by instruments alone. By the end of World War II, Link's company had sold 10,000 Blue Boxes to the armed forces (the name came from the standard color the Army painted them). The wartime need to quickly train hundreds of green recruits on a variety of airplanes made the Link trainer a serious tool for pilots (figure 8.7).

Link trainers used electrical and mechanical analog devices to simulate flight. The Whirlwind digital computer was intended to be used as a flight simulator, but never was. Still, the notion of using a digital computer running at real-time speeds

8.7

A trainee sits in one of Edwin Link's trainers.

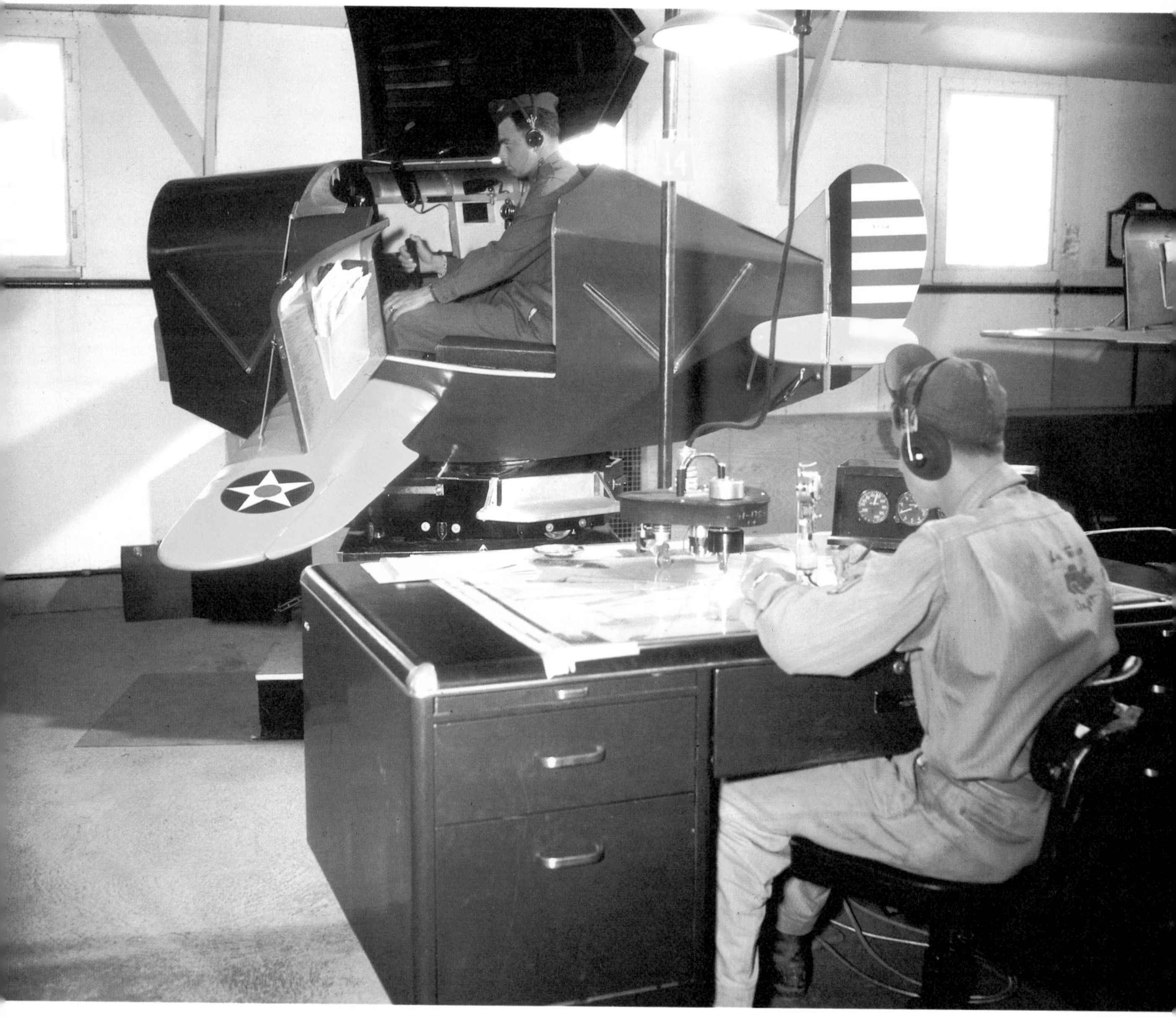
14

survived. As in flight testing, it took a while for the computer industry to come up with a machine that had sufficient processing speed. Before about 1970 most simulators still used analog devices. The Apollo program, whose goal was to land a man on the moon, put pressure on simulator manufacturers to provide more and more faithful simulations of spaceflight. This push from NASA dovetailed with the development by the computer industry of the minicomputer, a small and compact machine that could be dedicated to a task like flight simulation at a reasonable cost. Apollo used simulators of all sizes and shapes for every phase of the flight. One was even built to train the Lunar Module astronauts to descend the last one hundred feet to the Moon's surface (figure 8.8). Apollo astronauts spent over half of their preparation time in different simulators for various phases of the mission. After a dramatic and successful flight into space and back, it was common to hear the astronauts describe their experience as, "It was just like the simulator!"[4] (See figure 8.9.)

One thing the Link trainers did not simulate was the view out the cockpit window. They were intended to teach blind instrument flying, and during a typical training session the trainee pulled an opaque hood over himself so that he could see nothing but the instrument panel. Simulating the outside view was more difficult than re-creating the instrument readings. It was even more difficult than re-creating the pitching and yawing motion of the real craft, which could be done with hydraulic pistons rocking the simulator along various axes. But for some phases of pilot training a realistic view would greatly enhance the realism, and thus the effectiveness, of the training session.

Again with a firm push from NASA, simulator manufacturers came up with a technique of building a realistic scale model of a piece of terrain, and then "flying" a television camera over it under the control of the trainee's actions on the simulator controls. The image would be projected in the cockpit, using a special lens that re-created the illusion of depth (figure 8.10). The model boards sometimes had amazing amounts of detail, especially if used to train, say, helicopter

8.8

Apollo astronauts spent over half their training time in simulators. This terrain model was used to practice descending the final one hundred feet to the moon's surface. (Photo: NASA)

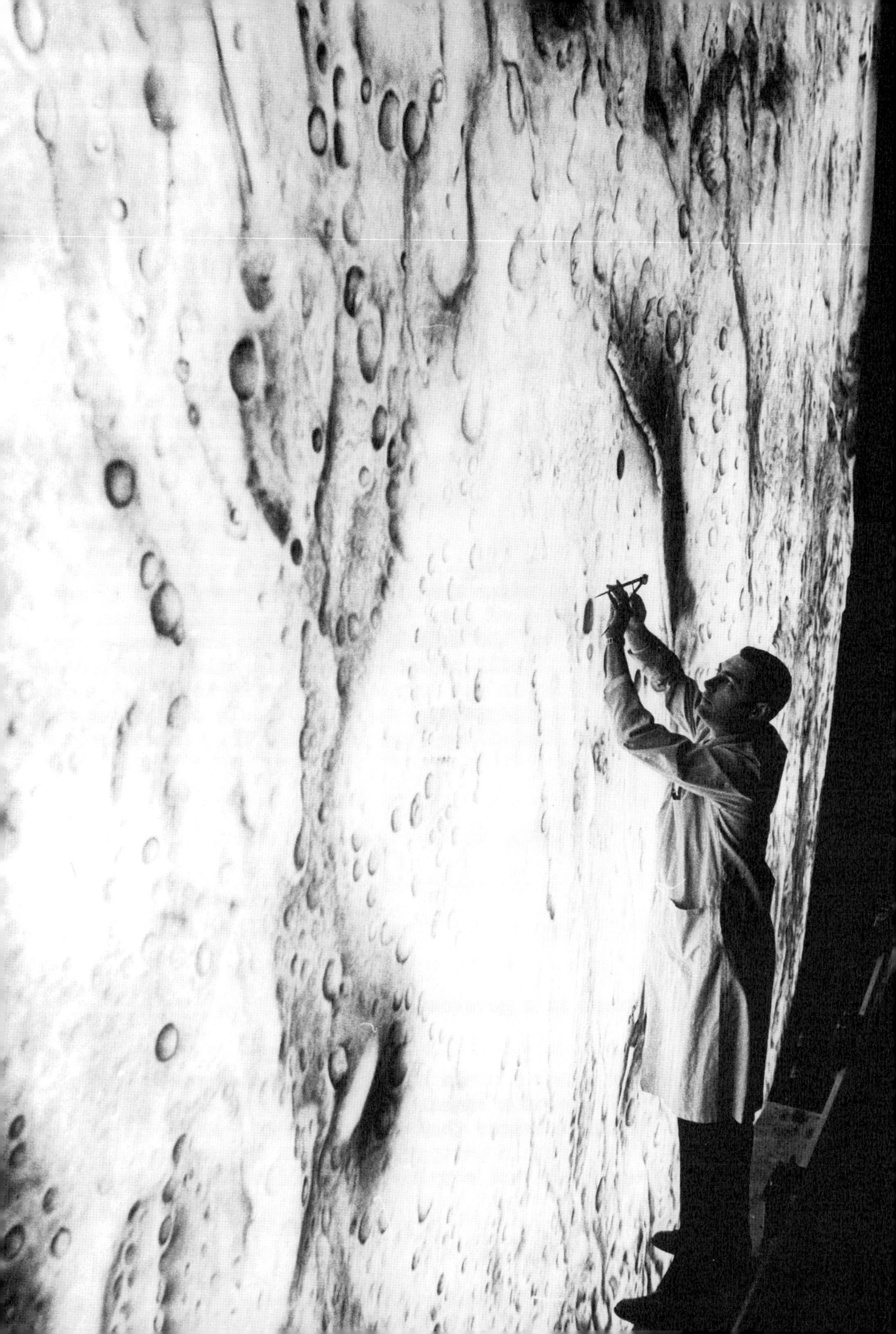

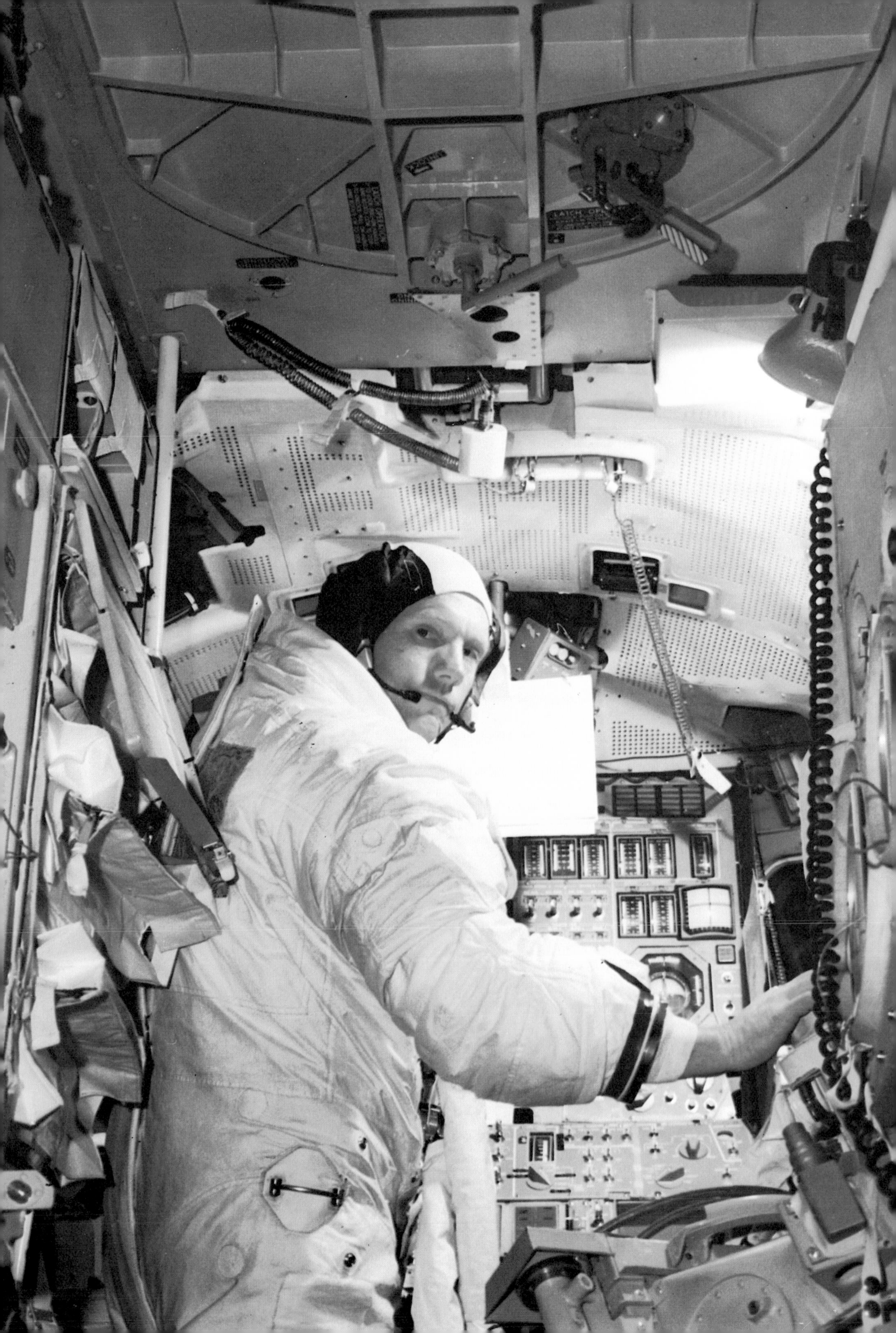

pilots in low-flying missions. For other missions such as attack or bombing runs, the boards had to be very extensive to simulate enough terrain.

By the late 1970s digital computers had taken over the task of generating these visual images as well. A visit to a video arcade will give an idea as to what computer-generated images look like; the professional systems used by airlines, NASA, and the military work on the same principle. The computer creates images of clouds, earth, sky, water, and buildings from mathematical descriptions of their basic shapes. As the trainee "flys" through a simulated space, the computer calculates what he should see in front of him and calls up the appropriate shapes from its memory (figure 8.11). To make all this seem real, speed is of the essence. So is the detail that the images must have: mountains must look rough and not like perfect pyramids; trees and grass must have the irregularities of nature to be convincing (figure 8.12). This is where home-computer simulators fall short. All these graphics require enormous computing power, equal to and even surpassing the power needed to run the rest of the simulator's functions.

The above examples of CAD/CAM, testing, and simulation show how the digital computer has penetrated into aerospace activity. They also show the extent to which computers allow one to move away from the actual nuts and bolts of flight hardware. Yet the refractory nature of flight requires that we abstract away from the hardware. At the simplest level the computer calculates the strength of pieces of metal, at the most abstract level it creates a virtual universe, in which a proposed aircraft design flies through a fictitious space, performing a hypothetical mission. At this extreme, nothing is real, and anything is possible. No wonder the computer has attracted the computer fanatic who cannot take his hands off the keyboard as he flys through a universe of his own making. But for aerospace and NASA engineers, all this has a serious purpose: to produce better performing air and space craft that human beings can fly and manage safely. By and large they do not succumb to the temptation to dwell in a realm of fantasy. Perhaps they do not because the reality of flying machines is fantastic enough all on its own.

8.9

Neil Armstrong, practicing a landing in the Lunar Module simulator. (Photo: NASA)

8.10

Before the development of real-time computer graphics, elaborate, detailed scale models of terrain were used to train pilots to fly various missions. The pilot's actions on the simulator controls activated a remote television camera, which "flew" over the terrain board just as a real aircraft might (*a*). This board, over sixty feet long, was used at the NASA Ames Research Center to train military pilots and Space Shuttle astronauts (*b*). (Photos: NASA)

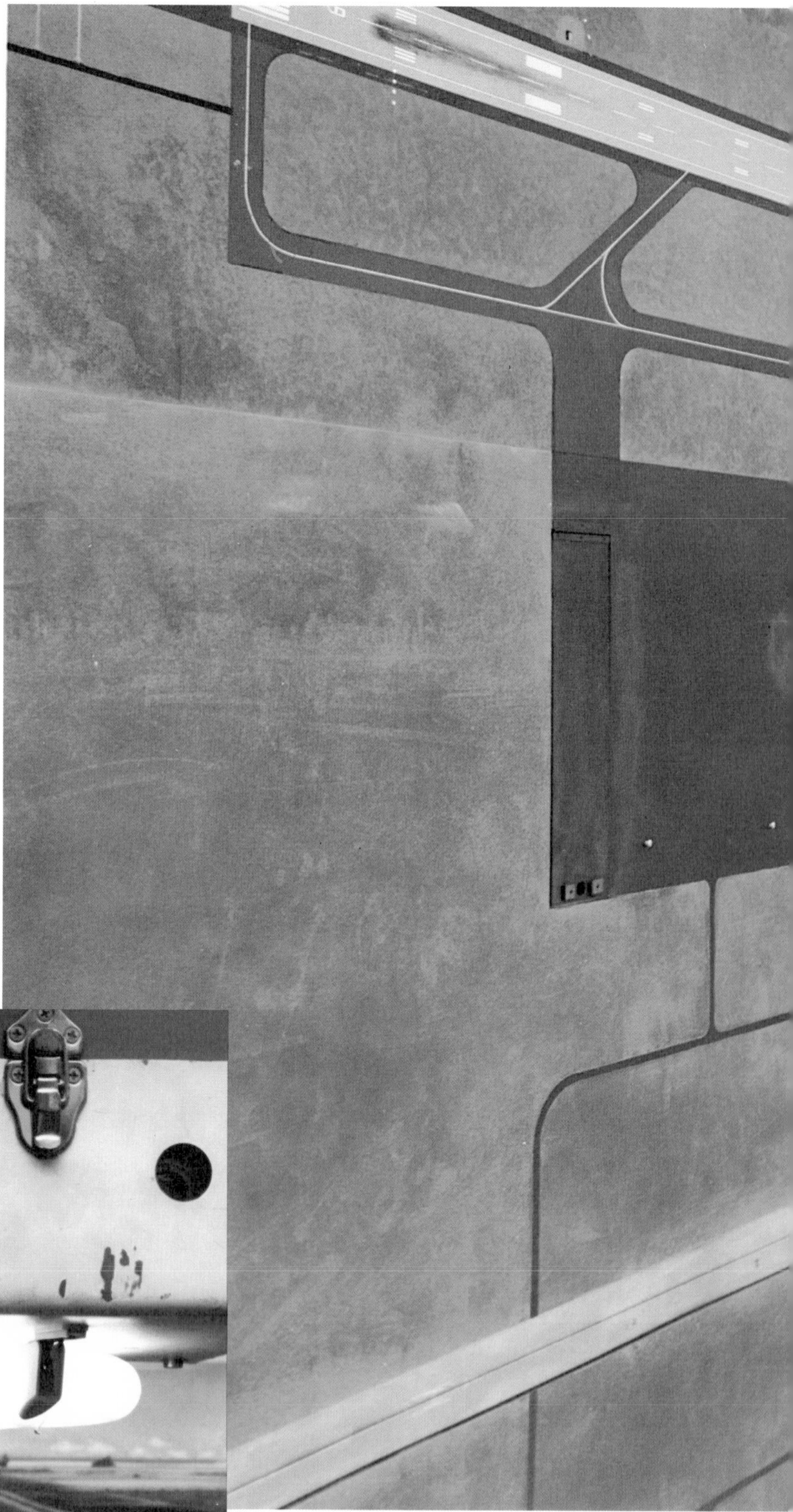

a

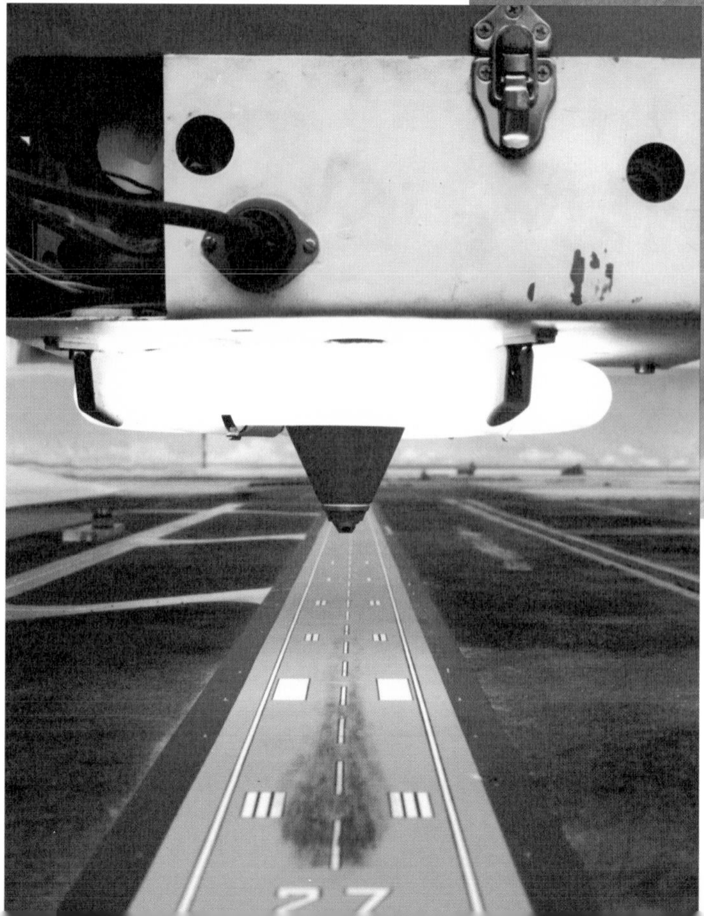

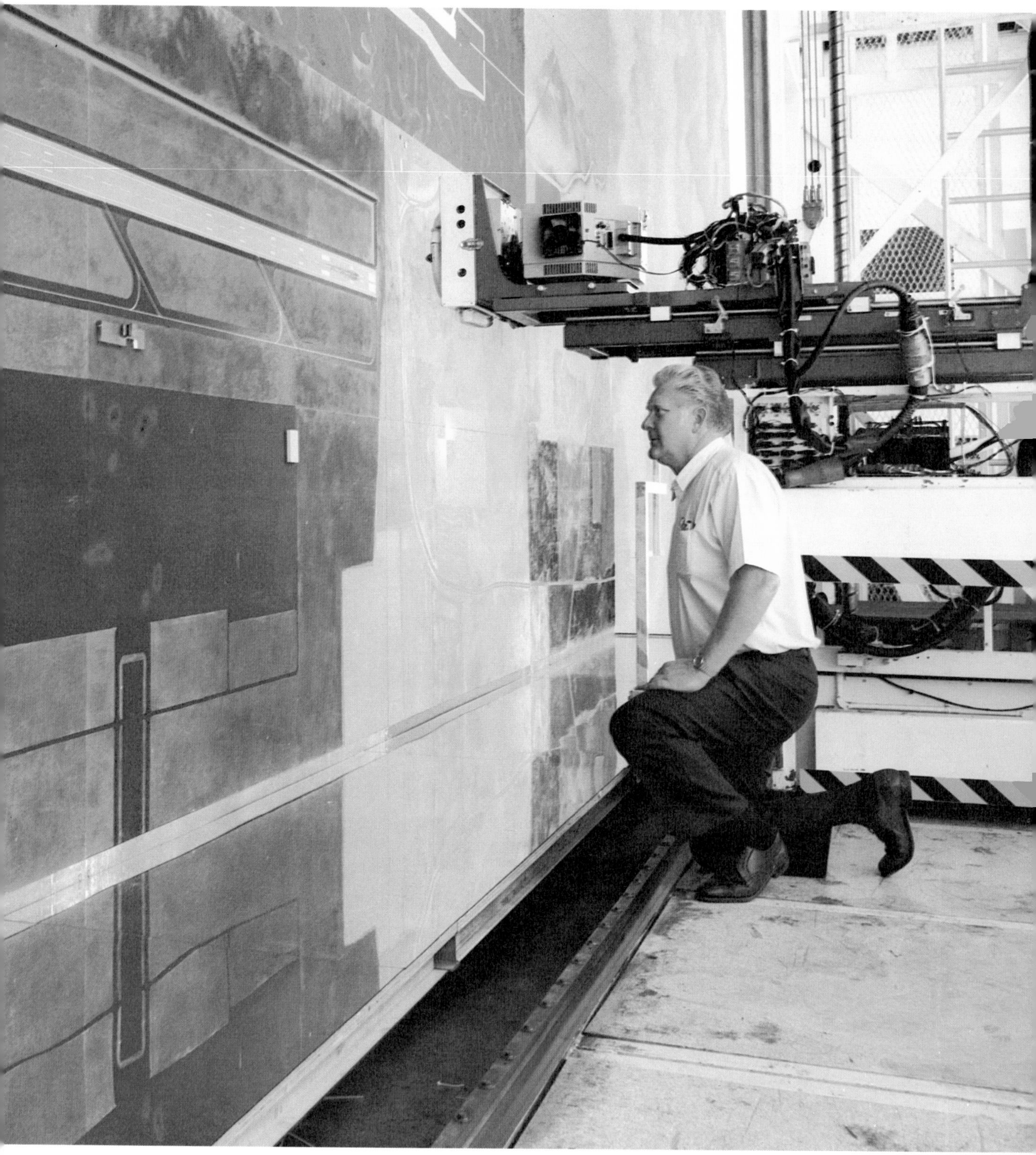

b

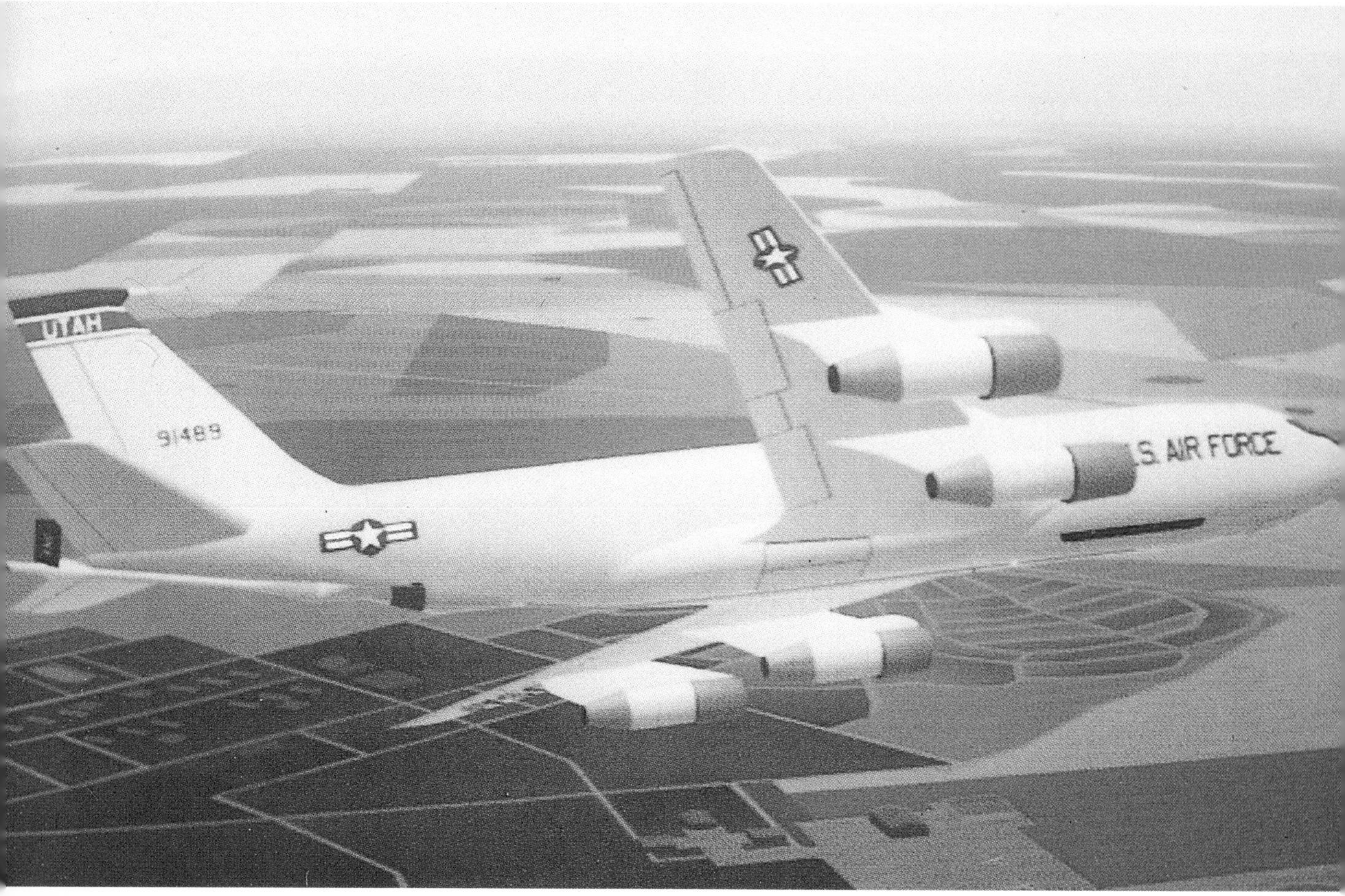
UTAH
91489
.S. AIR FORCE

8.11

8.11

Computer-generated images have now replaced terrain boards to recreate what the trainee sees out the window of the simulator. This image of an Air Force tanker is used to practice midair refueling. Note the realism of both the tanker and the surrounding terrain. (Photo: Evans and Sutherland Corporation)

8.12

Apache Helicopter is a popular game among home computer enthusiasts. Such games are remarkably accurate in their simulation of the behavior of real aircraft. They lack the rich visual detail of commercial and military simulators, mainly because of the limitations of personal computers. (Photo: Microprose)

8.12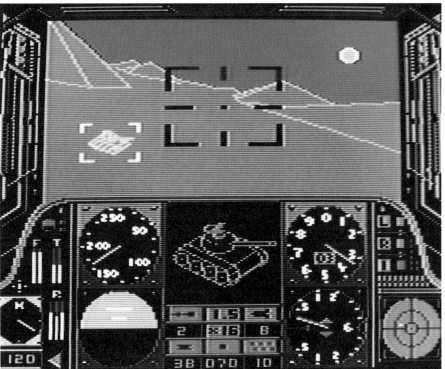

8.13

An air or space craft is free to move in any of six different ways. Three are rotations about its center of gravity: roll, pitch, and yaw (*a*). The other three are translations in space: speed (forward and backward), heading (left and right), and rate of climb (up and down) (*b*). In an airplane rotations and translations are intimately coupled. For example, to make a left turn, it is necessary to roll the plane to the left so that the wings will generate lift in that direction. But this is not so with a spacecraft, which operates in the absence of an atmosphere. For example, while in orbit, the Space Shuttle flies with its nose pointed forward, backward, up, or down according to its mission (e.g., taking a picture of the earth, pointing a telescope at a distant star).

a
Vertical axis
Positive pitch: nose up
Longitudinal axis
Lateral axis
Positive roll: to the right
Positive yaw: right turn
b
Vertical axis
Lateral motion
Longitudinal axis
Sideways
Lateral axis
Longitudinal motion
Forward
Vertical motion
Up or down

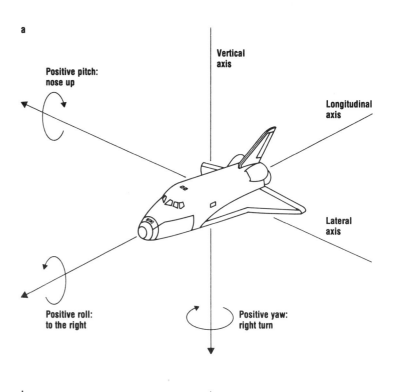

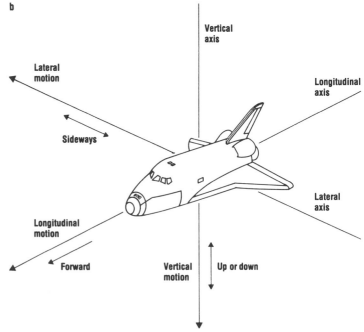

On-Board Control

The very quality that makes powered flight fantastic is also what makes it so difficult to master. That is its freedom: an airplane or spacecraft is free to move up, down, left, right, and forward at different speeds and at a variety of attitudes. It has six degrees of freedom: three describing its movement through space, three describing its attitude at any given position (figure 8.13). Contrast that with a railroad train, which has one: forward or backward; or an automobile, which has two: forward or backward, and left or right. (Automobiles have a very slight attitude freedom that depends on the geometry of their steering and suspension, but this is not controllable by the driver.) The digital computer is the latest in a long line of mechanical aids to control, but its flexibility and power have brought control to a new level of sophistication and complexity.

Safe flying requires a mastery of all six degrees of freedom. Managing that freedom was the greatest challenge to the pioneers of flight, and its mastery was what set the Wright brothers apart from their contemporaries who sought to conquer the skies. After extensive testing and experimentation with kites, gliders, and wind-tunnel models, the Wrights succeeded in achieving controlled flight, primarily by warping the wings to maintain lateral control. They also connected the cables that twisted the wings to the vertical tail fins, to coordinate these two control mechanisms during a turn (figure 8.14). It worked, but the Wrights' airplanes were by no means easy to fly.[5] As airplane design spread from the Wrights to others, the control problem remained an urgent one, and a better solution was sought.

The idea of using a spinning gyroscope to stabilize an aircraft occurred to some flyers not long after the Wright brothers' first flights. Gyroscopes had already been used to stabilize ships at sea, and their properties were well known. It would not do simply to mount a large and heavy ship's stabilizer on an airplane: the weight needed to have any stabilizing effect would also ensure that the plane never got off the ground! However, if a gyroscope's tendency to hold a fixed position could be used to supply *information* about the position of a craft, then it could form the heart of a system that augments the pilot's actions.

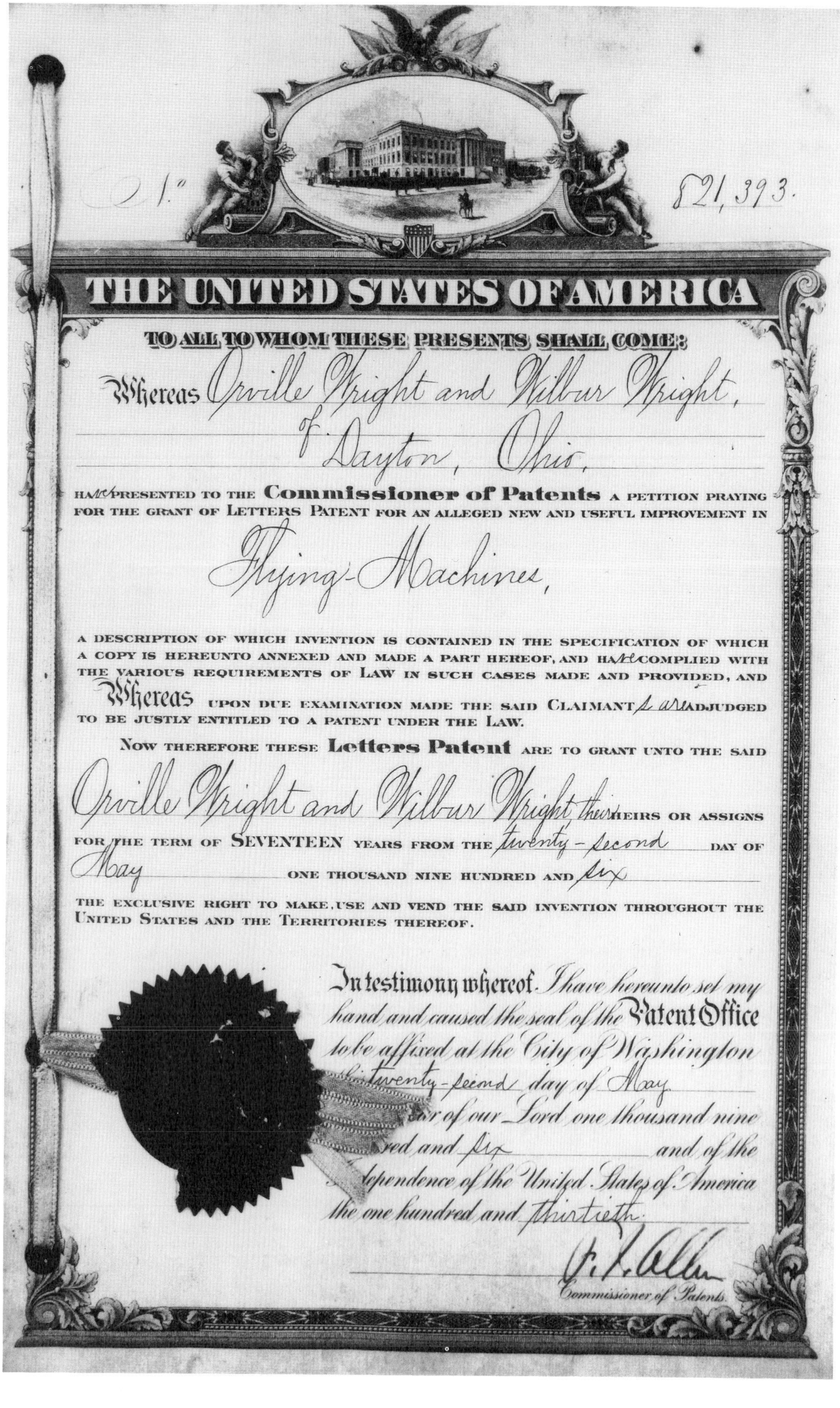

No. 821,393.

THE UNITED STATES OF AMERICA

TO ALL TO WHOM THESE PRESENTS SHALL COME:

Whereas Orville Wright and Wilbur Wright, of Dayton, Ohio,

have presented to the **Commissioner of Patents** a petition praying for the grant of Letters Patent for an alleged new and useful improvement in

Flying-Machines,

a description of which invention is contained in the specification of which a copy is hereunto annexed and made a part hereof, and have complied with the various requirements of Law in such cases made and provided, and

Whereas upon due examination made the said Claimants are adjudged to be justly entitled to a patent under the Law.

Now therefore these **Letters Patent** are to grant unto the said Orville Wright and Wilbur Wright their heirs or assigns for the term of SEVENTEEN years from the twenty-second day of May one thousand nine hundred and six the exclusive right to make, use and vend the said invention throughout the United States and the Territories thereof.

In testimony whereof I have hereunto set my hand and caused the seal of the Patent Office to be affixed at the City of Washington this twenty-second day of May in the year of our Lord one thousand nine hundred and six and of the Independence of the United States of America the one hundred and thirtieth.

F. I. Allen
Commissioner of Patents.

8.14

The Wright brothers patented their method of controlling a plane by warping the wings. They were the first to understand and solve the complex problems of stability and control of a powered aircraft.

No. 821,393.
PATENTED MAY 22, 1906.
O. & W. WRIGHT.
FLYING MACHINE.
APPLICATION FILED MAR. 23, 1903.
3 SHEETS—SHEET 1.
FIG. 1.
WITNESSES:
INVENTORS.
Orville Wright
Wilbur Wright
BY
ATTORNEY.

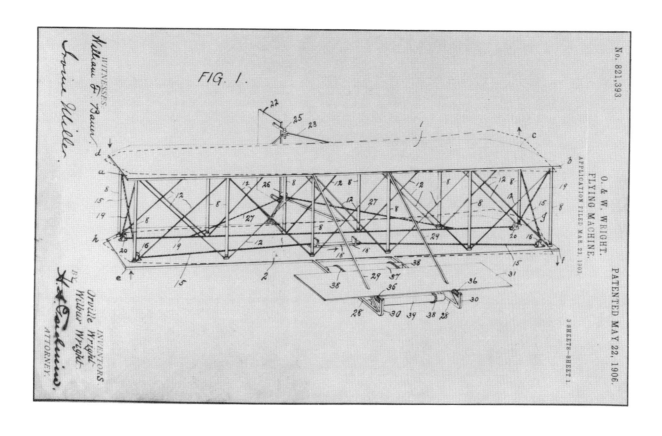

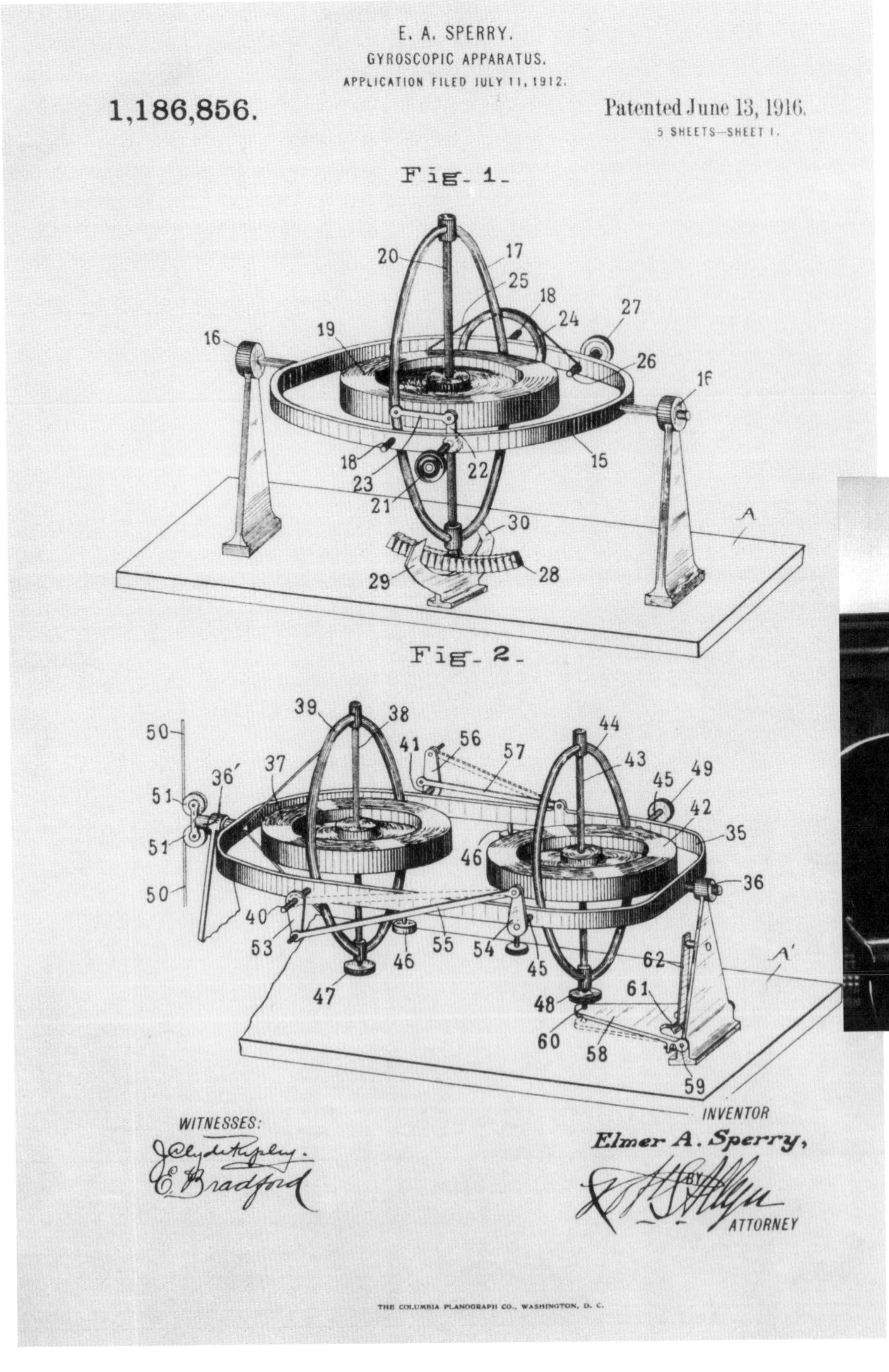
E. A. SPERRY.
GYROSCOPIC APPARATUS.
APPLICATION FILED JULY 11, 1912.
1,186,856.
Patented June 13, 1916.
5 SHEETS—SHEET 1.
Fig. 1.
Fig. 2.
WITNESSES:
INVENTOR
Elmer A. Sperry,
BY
ATTORNEY
THE COLUMBIA PLANOGRAPH CO., WASHINGTON, D. C.

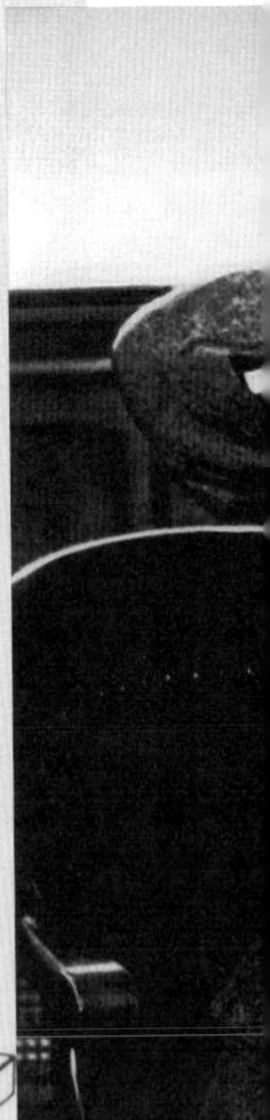

8.15

Elmer Sperry, ca. 1910 (*a*). A page from Sperry's patent for an airplane stabilizer (*b*). (Photo: Sperry Commercial Flight Systems)

a

That notion occurred to Elmer Sperry (1860–1930), a gifted inventor who had already developed gyroscopic systems for ships.[6] Working with his assistant Hannibal Ford, Sperry designed a system containing four gyroscopes that maintained stability along all three axes of freedom and that could stabilize an aircraft through electric and pneumatic links to the control surfaces of the plane (figure 8.15). Sperry's insight was to use gyroscopes not as active stabilization devices (as ships used them) but to create an artificial bedrock on board the plane, an Archimedean fulcrum against which to move the world. Sperry's system since became the basis for all autopilots and inertial-guidance systems. Such a system is now known as a stable platform. In 1914, Elmer's son Lawrence Sperry gave a dramatic public demonstration of the system at Bezons, France, where the Aero Club sponsored a contest aimed at increasing flight safety. Piloting a Curtis flying boat with his father's system installed, the younger Sperry stood up in the cockpit with his hands off the controls, while his assistant climbed out on one of the wings. The plane remained stable as it flew over the crowd (figure 8.16).

By the mid-1930s autopilots were in general use, and their capabilities were proved in a number of long-distance flights. Assisted by a Sperry A-2 autopilot, Wiley Post flew his Lockheed Vega, the famed *Winnie Mae,* solo around the world in 1933, the year that saw the commercial introduction of these systems (figure 8.17). The principle of gyroscopic stabilization led to one of the most famous secret weapons of World War II, the Norden Bombsight. This sight contained two parts, each with its own gyroscopes and electromechanical analog computers. During a bombing run one part computed the range and time to the target, while the other took control from the pilot and held the plane steady as the bombs were released (figures 8.18 and 8.19).

The function of an autopilot is to give an extra measure of stability and control to an airplane. Its need arose from the irony described above, namely, that the very freedom of motion that an airplane offers is also what makes it so difficult to master and safely fly. The Wright brothers' airplanes were controllable, but just barely so. They lacked an autopilot and were inherently unstable along both longitudinal and lateral axes. That meant they required constant attention from the pilot to

8.16

keep from losing control. Some of the Wrights' contemporaries argued that no one could fly such an unstable aircraft. But the Wrights were experienced bicycle mechanics. They knew that just because a vehicle is inherently unstable, it did not mean that a person could not control it.[7]

Just the same, after their success in demonstrating controlled flight in a series of flights between 1903 and 1908, other aircraft designers sought ways to build inherently stable craft making fewer demands on the pilot's skill and attention. By the 1930s the wisdom of aircraft design was that a plane should be inherently stable: that is, if disturbed by a gust of wind or the like, it should return to its original position of pitch, roll, and yaw.[8] The pilot will still need to keep his hands on the controls, but only to direct the plane's heading, speed, and altitude, not to keep it from pitching over.

Designing a plane to act this way over a wide range of speeds was a complex problem, involving much theoretical analysis, wind tunnel testing, and flight testing of prototypes. To stabilize a craft's pitching motions, designers came up with the

8.16

Lawrence B. Sperry soloing in a Curtis pusher biplane at Hammondsport, New York, ca. 1912. (Photo: Sperry Commercial Flight Systems)

8.17

Wiley Post used this Sperry A-2 Autopilot to help him make a solo round-the-world flight in 1933. (Photo: David E. Dempster)

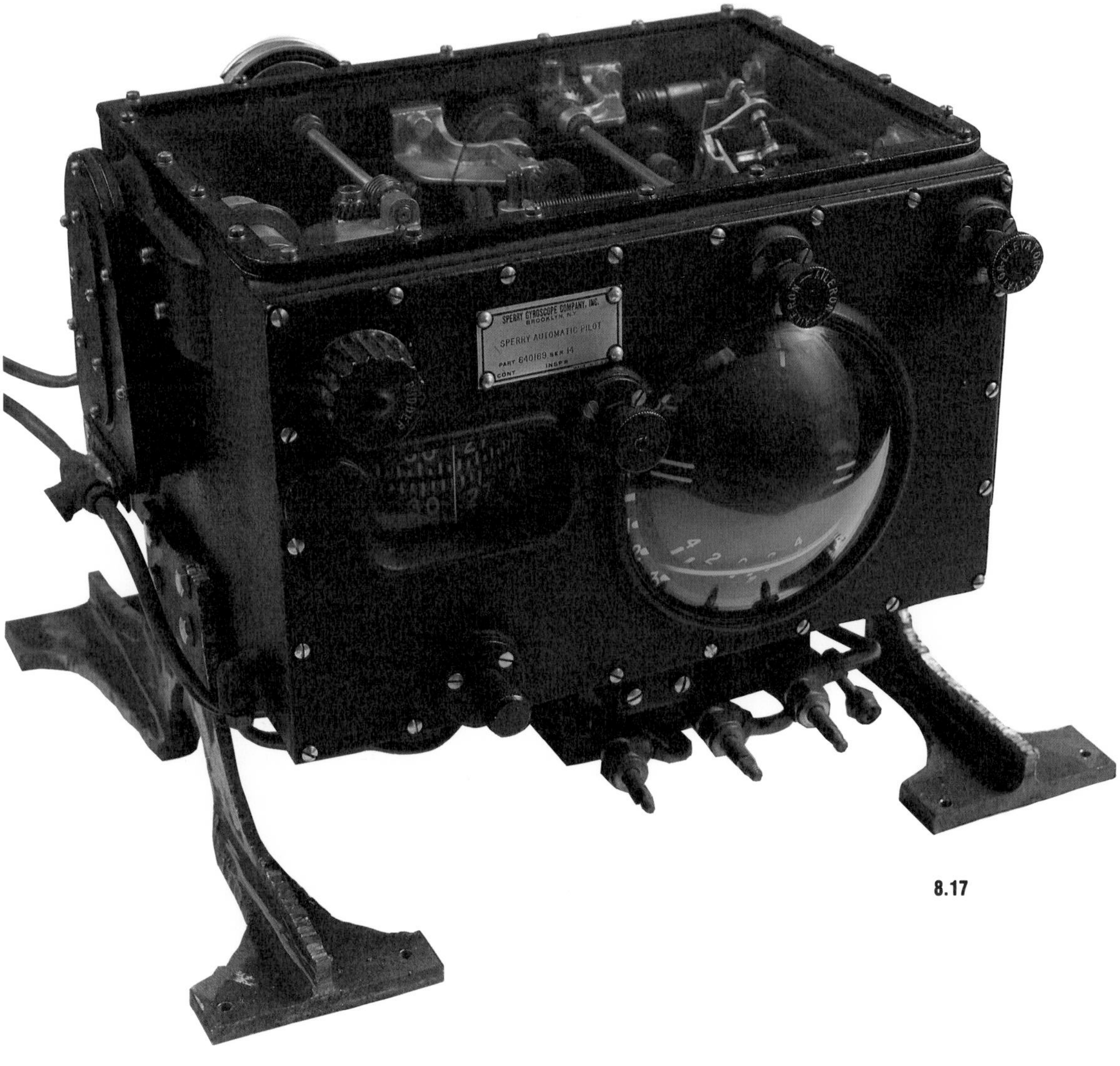
SPERRY GYROSCOPE COMPANY, INC.
SPERRY AUTOMATIC PILOT

8.17

8.18

This is the prototype for the Norden Bombsight, built in 1923. During World War II thousands were used on U.S.-built bombers. (Photo: David E. Dempster)

8.19

Pilots and crew relied on numerous special-purpose analog slide rules to assist them in flying and navigating. These are just a few of the hundreds of rules available. Many of these functions are now performed more cheaply and accurately by on-board digital equipment. (Photo: David E. Dempster)

8.18

FOR SECTIONAL CHARTS
WARNER PLOTTER-COMPUTER FDF-16
MFG. BY Felsenthal INSTRUMENTS CO.
NUCLEAR BOMB EFFECTS COMPUTER
GROUND SPEED – M.P.H.
MINUTES

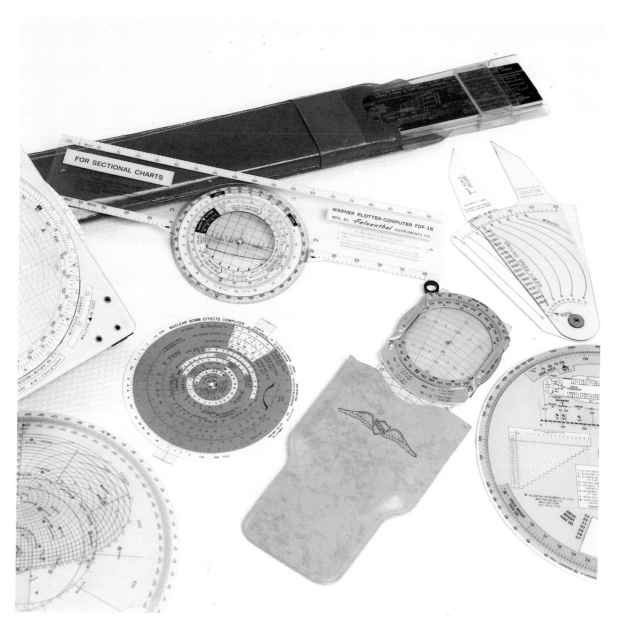

8.19

solution of placing the center of lifting force of the wings slightly behind the plane's center of gravity and then having the horizontal stabilizer exert a downward force to counteract any tendency to pitch over (figure 8.20). Note that the horizontal stabilizer is mounted in the rear of the plane, not in the front, where the Wrights placed it. This placement does not automatically make the plane inherently stable, but it does make it easier to design stability into a plane, just as tail feathers make an arrow fly straight. (Nor does forward placement automatically make the plane unstable, but in the Wrights' plane it did.) The next section will deal with this configuration in greater detail. This and the following sections deal with stabilizing an airplane's pitch; the basic principles also apply to stabilizing roll and yaw.

The digital computer, coupled with precise gyroscopic platforms, changed this picture. From the U.S. manned space program, aircraft designers borrowed the notion of using an on-board computer to continuously monitor the position of an airplane and continuously adjust its control surfaces to keep it stable. The computer, drawing information from a gyro-based stable platform and from engine and airspeed sensors, replaced the ever-vigilant pilot of the Wright brothers' era. That allows designers once again to design craft that are inherently unstable and hence much more agile. This idea of active digital controls conferring stability on an inherently unstable design is the chief legacy of the space program on modern aeronautics. The idea came from spacecraft designers, who realized stability *has* to be artificially created in the vacuum of space, since there are no gravitational or aerodynamic forces to work with.

HiMAT and X-29

Relaxed stability and active computer control are already used in the F-16 and F-18 fighter planes, and they are crucial to the ability of the Space Shuttle to reenter the earth's atmosphere and glide to an unpowered but controlled landing on earth. New-generation commercial jets (like the Airbus A-320) are also using these systems. The concept is best illustrated by two NASA research aircraft already mentioned: the HiMAT

8.20

An airplane designed to be inherently stable returns naturally to a position of equilibrium without any action by the pilot (*a*). The action may be compared to a ball placed in a shallow dish or a flat plate (*b*). The Wright brothers' planes were unstable and required constant attention from the pilot to keep them on a straight and level path. Over the years aircraft design has tended toward more and more inherent stability, mainly for reasons of safety. But in recent years there has been a return to inherently unstable designs (like the X-29), because such designs offer greater maneuverability.

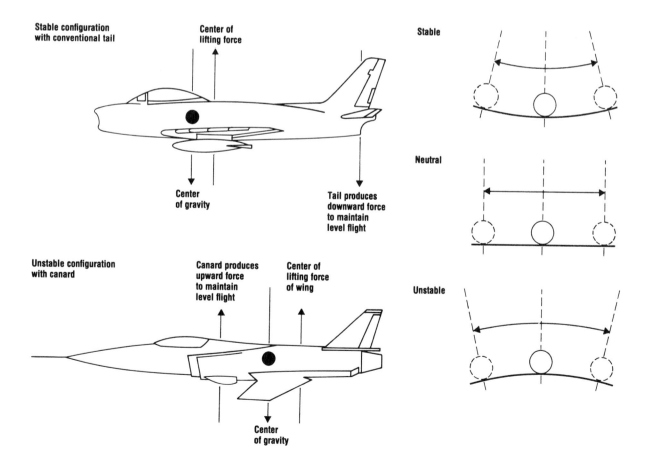

Stable configuration with conventional tail

Center of lifting force

Center of gravity

Tail produces downward force to maintain level flight

Unstable configuration with canard

Canard produces upward force to maintain level flight

Center of lifting force of wing

Center of gravity

Stable

Neutral

Unstable

and the X-29. Earlier chapters described how computers helped design and test these planes, but their on-board computer controls set them apart from ordinary craft even more.

Both craft have their horizontal stabilizer forward of the wing, where the Wright brothers had it. There it is called a canard, probably because a forward-mounted stabilizer resembles a duck's bill. On these two planes the canard helps stabilize the plane by providing additional lift, instead of a providing a downward force as a conventional tail does.[9] This allows the main wing to be smaller, since the canard provides some lift, and the main wing does not have to generate lift to compensate for the tail. And this in turn means a lighter plane, which thus accelerates faster and maneuvers more quickly.

It is possible to have a canard configuration that is inherently stable, but on the X-29 the wings, canard, and center of gravity are configured in a way that makes the plane inherently unstable. This configuration means that the plane offers less inherent resistance to violent turning and banking, just what is needed in an aerial dogfight. To keep the plane from tumbling, a digital computer steps in and actively adds a measure of artificial stability. It does this by constantly sensing the attitude of the plane and adjusting the canard and the flaps on the wing accordingly. The digital computers on the X-29 make these adjustments forty times a second, much faster than any human pilot's reaction time. These corrections take place so rapidly and smoothly that the pilot is unaware of them. In other words, with the help of a digital computer, the X-29 flies just like an airplane that is inherently stable (figure 8.20).

Note that this feel of stability is provided by the computers and can thus be tailored by changing the computers' programs. It is even possible to program the X-29's computers to make it feel inherently unstable, if its designers and pilots wanted to. (This sense of agility and stability is similar to the different steering feel of a race car in contrast to a family sedan.) Finding the optimal amount of artificial stability that pilots like and that keeps the X-29 within a safe envelope is one of the goals of the extensive test program now underway at NASA's Dryden Research Center in California. Especially interesting here is that a major portion of this test program involves changing the software of the X-29, not its hardware (computer or other hardware).

These systems maintain static stability in the three positional axes of pitch, roll, and yaw. The same systems can also provide control over the translation motion of the plane: when the plane climbs or descends, goes faster or slower, or changes heading. Using the computer in this way is now gaining acceptance. It is the essence of a new generation of "fly-by-wire" airplanes and spacecraft.

Fly-by-Wire Control

A pilot makes a simple turn by moving both the ailerons (located on the trailing edge of the wings) and the rudder. The ailerons roll the plane to one side, which produces some lift on that side, while the rudder points the nose in the direction of the turn. Because some of the wing lift is now helping turn the plane, there is less available to keep its altitude, so the pilot must also advance the engine throttle. A smooth turn requires coordinating these three actions.

The earliest airplanes connected the stick and rudder pedals directly to the ailerons and rudder with piano wire and pulleys. As aircraft got heavier and more complex, some assistance became necessary, usually from hydraulic systems similar to the power steering and power brakes in an automobile. As in automobile steering, the hydraulic system meant that the feel of the stick and rudder did not correspond exactly to their effects on the plane's control surfaces. But there still was a direct correlation between what the pilot did and how those surfaces behaved. Again like automobile steering, if hydraulic pressure were lost, the pilot could still move those surfaces, though only with a lot more muscle power.

In a fly-by-wire control system the pilot's actions on the controls go to a digital computer. The computer's memory stores flight-test, wind-tunnel, and engineering data on the performance of the plane, and from a set of on-board sensors it collects data on the current altitude, airspeed, position, and engine power. From all these factors the computer can then determine just what action or combination of actions is needed on the flaps, rudder, throttle, and the like to effect whatever the pilot wants to do. Once again, all this happens so fast that the pilot is unaware that a computer is in the loop.

Why do things this way, which is after all much more complicated? Consider again current automobile technology. If a driver applies the brakes on an icy road, the wheels will lock and the car will skid. Power brakes only make this situation worse. But some new automobiles have a microprocessor in the brake system that senses when a wheel is locking up and then lessens the brake pressure until the wheel regains traction. It does all this regardless of how much pressure the driver puts on the brake pedal and also much faster than a human being could sense and react anyway. The driver still feels the car responding to his desire to stop, but the brakes no longer directly respond to his commands. The same holds true with fly-by-wire airplanes. Such a plane's on-board computer intervenes between the pilot and the control surfaces and carries out the pilot's wishes in the best way, one that the pilot may not know of or be able to perform. Obvious cases are to prevent a maneuver of such violence that it would cause structural damage to the airplane or a maneuver that might cause it to stall and crash. For a commercial transport, the computer can minimize fuel consumption while maintaining a comfortable ride for the passengers.

Redundancy

The biggest drawback to fly-by-wire control is that the computer must work if the plane is to fly at all. There is no manual connection, as there is in hydraulically assisted controls. Like other machines, computers can fail. But a plane designed to utilize the computer's power cannot have a manual override or backup. By design such a craft is unflyable by an unassisted human being. Many subsystems of an airplane or spacecraft have no backups—the wings, for example. But a designer tries to minimize the number of these as much as possible. The risks of failure in a fly-by-wire system are serious and far greater than those for a conventional control system for which a manual backup is possible. Why, then, is it adopted?

As it happened, fly-by-wire technology found its first application when there was no other choice but to use it. The first flying machine to have a computer play such an active role was not an airplane but a spaceship: the Lunar Module, which carried two Apollo astronauts from lunar orbit to the moon's surface and back in a series of missions between 1969 and 1972. Landing on the moon was too complicated a job for a human being to perform unassisted, and the distance from earth meant that they could not rely on earth-based computers for real-time instructions. So the landing was handled by a digital computer identical to the one in the Command Module (see chapter 6). Whereas the Command Module's computer was programmed to handle guidance and navigation through space, the Lunar Module's computer was programmed to assist in the lunar landing, the ascent from the lunar surface, and rendezvous and docking with the Command Module.

The Apollo Guidance Computer had no redundancy in it. Instead, it was designed, built, and tested to extremely rigid specifications to ensure that it worked right the first time. If the Command Module's computer failed, a combination of commands from the ground and manual navigation aids on board could get the astronauts back to earth. If the Lunar Module's computer failed, a backup Abort Guidance Computer would do the minimum required to fire the ascent rockets and return the ship to lunar orbit. But it could do nothing more. During manned Apollo missions the Lunar Module computers never failed. The Command Module computer failed once: an explosion on the Apollo 13 Service Module knocked out all power to the Command Module and its computer as well. The astronauts safely navigated back to earth by using commands from Houston and by using the Lunar Module and its computer as a lifeboat of sorts.

The success of Apollo led to the transfer of fly-by-wire control first to aircraft and then back to spacecraft, namely, the Shuttle. In 1970, as the Apollo Program was winding down, a group of NASA engineers installed an Apollo Guidance Computer in an F-8 airplane. They sought to test continuous digital control in an aeronautic environment in anticipation of designing a space shuttle that could be flown to a controlled landing on earth.

Although the software requirements were vastly different, the Apollo computer had proved reliable, and because several lunar missions had been canceled, there were several surplus computers available. Since the computer was a general-purpose digital computer, it was not that difficult to program it for an aeronautic application. Indeed, the difference between the F-8 and the Lunar Module missions was comparable to the difference between the Apollo Command and Lunar Module missions. The MIT Instrumentation Lab (now called Draper Labs) developed the flight software, as it did for the lunar missions. The F-8 made its first digitally controlled flight in May 1972 at NASA's Dryden Research Center.[10]

The F-8 tests produced results that convinced Shuttle designers that active digital controls could handle the Shuttle as it traveled from the vacuum of space, through the heat of atmospheric reentry, to an unpowered landing on an ordinary runway. It was not feasible to continue using surplus Apollo computer hardware for the Shuttle, and the cost of making that computer reliable was so high that Shuttle designers adopted an alternate system, which has since become standard for current fly-by-wire designs. They fitted the craft with several copies of the same computer, each running the same program and continuously comparing their output.

This concept requires a minimum of three identical computers to work; the Space Shuttle has five, of which four form a redundant set and one functions as a separate backup. If one computer fails in the Shuttle, it is outvoted by the others and the mission can continue with no danger to the crew. If another fails, the remaining three still form a viable system. Any one computer can bring the Shuttle back to earth safely if it is working properly. The odds of multiple, simultaneous failures are remote enough to make the Shuttle's controls one of its most trustworthy systems. (From the Shuttle's first flight in 1981 to the suspension of flights after the *Challenger* explosion in 1985, there have been a few instances in which one computer failed. In one case, the Spacelab 1 Mission of December 1983, one computer then a second failed just prior to reentry; so did one of the three redundant inertial-measurement units on which the computers depended for attitude information. The remaining three computers allowed Pilots John Young and Brewster Shaw to land safely. The failures were later found to be due to bits of dirt contaminating the integrated circuits.[11])

The X-29 has three identical digital computers (made by Honeywell) and an analog backup in case of multiple failure. The Shuttle's computers are based on a standard design built by IBM for manned bombers. NASA chose the IBM AP-101 because it was a proven design and had a long track record of reliable operation. By modern standards these computers are obsolete, as is the use of redundancy to achieve reliability. In the past five years a number of computer companies have introduced highly reliable commercial products that use redundancy, but at the circuit-board level, rather than at the level of complete systems, as in the Shuttle and X-29. This approach offers the same reliability while reducing weight and giving higher processing speeds. But the proven track record of the IBM and Honeywell avionics computers has made NASA reluctant to try this new approach. The next generation of spacecraft and fly-by-wire airplanes will no doubt incorporate these new computers.

Redundancy has solved the problem of how to make fly-by-wire controls reliable. Redundant computers give the same reliability to controls that multiple engines give to commercial passenger planes. But there is one difference, the software. Even if all the hardware is functioning normally, an undetected flaw in the computer programs could cause a crash. In the Shuttle, four of the computers have identical software, but the fifth is deliberately programmed by a different team to guard against a common software error. (IBM's Federal Systems Division, which supplied the Shuttle's computers, also does most of the programming; Rockwell International, which builds the Shuttle Orbiter, programs the fifth computer.) So far, no common software errors have required the use of the backup program. But the fifth computer and its different software are only a fallback; true software reliability has to be achieved by the same rigor in design, production, and testing as with any other piece of an airplane or spacecraft. How aerospace engineers learned this lesson is the subject of the next chapter.

Chapter 9
Software

"Software" is the name given to the procedures and programs that cause a computer to do what it does. Because a computer is by nature a universal machine, it requires a substantial body of such instructions to give it its real purpose; otherwise, it cannot realize its potential. The common notion of software is that it is a set of commands and instructions to a computer, written out in a code, or programming language, and interpreted by the computer's circuits. While this is correct, it does not capture the nature of programming a computer. Programming is not so much an act of giving instructions as it is an act of designing a system that performs some specific task. The hardware, a general-purpose digital computer, is the substrate on which the programmer builds his machine. Only with the addition of software is that machine complete.[1]

The word "software" implies a distinction from traditional engineering hardware, but it tells us little beyond that.[2] Software sometimes has a physical form: magnetic tape, disks, or in an earlier day, punched cards. These media are often physically soft enough that one can bend them; others may be physically hard and durable. But these media are not software: software is the information they carry; as such it has no mass or volume. Software is usually associated with programming languages like Fortran, BASIC, or APT; but it encompasses any of the ways one can encode instructions so that a machine can follow them. These range from individual commands encoded in binary numbers to applications packages for personal computers for word processing, spreadsheet calculations, and games.

In a computer the software is stored in the memory unit along with other data, and it is from this location that it does its work. So the word "software" has less to do with its physical form than with its changeability and intimate connection with the system's purpose, supplied by the computer user. Software techniques have not advanced at the same astonishing rate as innovations in microelectronics hardware since the 1950s, but only a combination of both has given us the computer revolution of the past two decades.

Programming a computer has much in common with other aspects of machine design, but few other engineers must handle the orders of magnitude handled by programmers, especially systems programmers. Those who direct the writing of computer programs that control pieces of the U.S. air-traffic-control system, for example, must know not only the general structure of that huge system but also something of how each individual instruction is carried out in the deepest recesses of the computer's circuits. Some programmers work long and hard at refining a program to save a few cycles of machine time on the order of a few nanoseconds. In a real-time system like the air-traffic-control system these nanoseconds can spell the difference between success and failure if the hardware is already being pushed to its limits.[3] Also, the programmer must be sure that every instruction is correct down to its last bit, or else the entire system may fail. The orders of magnitude involved in this work are many times those found in any other engineering discipline.[4] Aerospace software shares these characteristics with software used in other computer applications, just as the hardware that aerospace engineers use is the same as that which other engineers use. But in some respects aerospace software is distinctive. Aerospace has played an important role in developing the following facets of computer performance, and since the mid-1970s other segments of society, especially the financial communities, have adopted them as well.

Earlier chapters emphasized the numerous aerospace computer applications that operated in real time. Aerospace software must therefore be written to operate in this way. Few of the standard programming languages and operating systems had this capability. Aerospace industries, beginning with the System Development Corporation's work on SAGE, brought real-time programming to digital computers.

Aerospace computer applications were the first to require large systems of multiple computers, radar stations, communications equipment, and remote terminals. Projects Hurricane and Whirlwind were the pioneers, and SAGE laid the groundwork for large networks such as the U.S. Air-traffic-control system and American Airlines' SABRE reservation system (the largest private computer network in the world). Also furnishing a basis for commercial systems was NASA's ground-control network for Apollo, which tied the Mission

Control Center in Houston with the launch pad at Cape
Canaveral, other tracking stations around the globe, and the
recovery ships at sea. These systems are growing throughout
the world. In nearly all cases they are descendants of those
early pioneering projects. Systems software is by nature far
more complex than other kinds of software.[5]

Finally, aerospace programmers were among the first to con-
front the problem of reliability head-on. A serious computer
malfunction may be annoying in a ground-based activity; in a
computer involved in flying a craft, it is intolerable. Advances
in microelectronics have produced hardware that operates
very reliably, but less software that runs so. Thus, most of the
story of aerospace software is about how one can make it reliable. What follows is not an exhaustive treatment of this story,
but rather some historical examples that illustrate the depth
and complexity of the issue of software reliability. Among the
many things these examples illustrate, one that stands out
is how software anomalies can occur at any level of detail,
from the global fit of the software with the outside world
down to an individual byte in an instruction coded deep
within a processor.[6]

Bugs

A program that does not do what one wants it to has a "bug"
or several "bugs" in it. Many believe the term was coined by
computer pioneers in the 1940s, but in fact Thomas Edison
used it when talking about an electric power system in the
1880s.[7] Today the word refers mainly to computer software
problems, but the word has no precise definition. Computer
errors come in a variety of types, and bugs can range from an
obviously wrong instruction to multiply instead of to add two
numbers together to a vagueness in a customer's specification
that results in a system that only vaguely does what was
wanted. The richness and complexity of a computer's behavior, coupled with its maddening refusal to "do what I mean
and not what I say," makes for a wide variety of surprises.
Debugging is a subtle and difficult art.

In a typical Shuttle Mission the astronauts refer to a book of
program notes that tell them of minor inconsistencies in the
mission software. The source of these notes is often unex-

pected or unusual system behavior uncovered during a simulation on the ground. Many of them are simply statements of what a programmer considers normal computer behavior, which astronauts are less comfortable with. At most they are minor inconveniences and are far from life-threatening.[8] But the fact that astronauts have to carry this book reveals how the best efforts of NASA, Rockwell International, and IBM to produce perfect software are at best elusive.

For the Shuttle program NASA established a procedure in which every line of program code is subject to review and testing at ever-increasing levels of integration. Only after passing the final level, which involves hours of simulated flying on the ground, is the software "man-rated." This procedure, pioneered by NASA, is patterned exactly after the design reviews of flight hardware for not only the Shuttle but all types of flying machines. It does not guarantee error-free software. But it does the job.

Nothing in flight engineering is certain. Many computer scientists, especially those in the universities, argue that error-free software is impossible. NASA, the Air Force, and aircraft manufacturers do not deny this belief but rather seek ways of reducing the risk to an acceptable level. These methods have taken on the name "software engineering," a discipline whose roots lie deep in aerospace and defense organizations and which has a much greater operational component than its sister discipline computer science.

The principles of software engineering are no different from other aspects of aerospace engineering, where the need for absolute strength and reliability must always be balanced against a need to reduce weight. But this comparison is hardly apt. Computer programs bear little resemblance to a wing spar or landing-gear strut, even if the programs are integrated into all the functions of the craft. Can software be engineered to be as reliable as other components of an airplane or spacecraft? Obviously it must be if the revolution in digital control described above is to succeed. But how can one speak of "engineering" something that by definition is not physical?[9]

Mariner I

In the first decade of electronic computing few realized that getting programs right was going to be a serious problem. An event in 1962 changed that. One might say it was the event that founded software engineering. On the morning of July 22 Mariner I, intended to be the first U.S. spacecraft to visit another planet (Venus), rocketed off the launch pad at Cape Canaveral atop an Atlas booster. After about four minutes into the flight, the Atlas booster began moving on an erratic and potentially dangerous course, and the range safety officer at Cape Canaveral ordered it destroyed. An investigation later found that the problem was due to a "hyphen" (actually a bar) inadvertently left out of the guidance equations used to write the program for the computer that monitored and directed the Atlas booster from the ground. This story, since repeated and embellished upon subsequent retelling, summed up the whole problem of software reliability. A missing hyphen became the most famous bug of them all.[10]

It is worth unraveling a bit of the story. During the launch the Atlas booster rocket was guided with the help of two radar systems. One, the Rate System, measured the velocity of the rocket as it ascended through the atmosphere. The other, the Track System, measured its distance and angle from a tracking antenna near the launch site. At the Cape a guidance computer processed these signals and sent control signals back to the tracking system, which in turn sent signals to the rocket. Its primary function was to ensure a proper separation from the Atlas booster and ignition of the Agena upper stage, which was to carry the Mariner Spacecraft to Venus.

Timing for the two radar systems was separated by a difference of forty-three milliseconds. To compensate, the computer was instructed to add forty-three milliseconds to the data from the Rate System during the launch. This action, which set both systems to the same sampling time base, required smoothed, or averaged, track data, obtained by an earlier computation, not the raw velocity data relayed directly from the track radar. The symbol for this smoothed data was $\dot{\bar{R}}_n$ or "R dot bar n," where R stands for the radius, the dot for the first derivative (i.e., the velocity), the bar for smoothed data, and n for the increment.

The bar was left out of the hand-written guidance equations.[11] Then during launch the on-board Rate System hardware failed. That in itself should not have jeopardized the mission, as the Track System radar was working and could have handled the ascent. But because of the missing bar in the guidance equations, the computer was processing the track data incorrectly.[12] The result was erroneous information that the velocity was fluctuating in an erratic and unpredictable manner, for which the computer tried to compensate by sending correction signals back to the rocket. In fact the rocket was ascending smoothly and needed no such correction. The result was *genuine* instead of phantom erratic behavior, which led the range safety officer to destroy the missile, and with it the Mariner spacecraft. Mariner I, its systems functioning normally, plunged into the Atlantic.

The Myth

This event has since become part of computer folklore and has taken on a mythical dimension. Several introductory college textbooks on computer programming cite it in their first chapters. Most of these accounts wrongly state that it was a simple typographical error that caused this failure. Other accounts have the basic facts wrong: they state that it was not a missing bar, but the substitution of a period for a comma in a Fortran program that caused the failure.[13] The example is cited to show, among other things, the need for careful typing, the computer's lack of common sense, and above all, the inadequacies of Fortran, which did not detect the error when the program was compiled.[14]

The fact that this story has been elevated to a central myth of software engineering illustrates several points: Correct software requires accurate typing, proofreading, and language syntax. Correct software also requires an accurate mathematical analysis of the problem. Finally, correct software requires the customer to have a clear knowledge of what he really wants the computer to do. That last requirement is the most difficult to achieve.

Stories like these appear frequently in the popular press, and make up a large part of the general public's perception of computers. Computer professionals too trade these stories. Most of the time they tell them to one another over computer networks, like the ARPANET set up by the Defense Advanced Research Projects Agency in the 1960s. Every few months the more interesting anecdotes are gathered off the electronic networks and published in a column called "Risks to the Public" in *Software Engineering Notes,* an informal publication of the Association for Computing Machinery. Most of them are related to aerospace. Like the Mariner I story, these stories are a mixture of a little bit of fact, a lot of hearsay, and sometimes a dose of embellishment. One can dismiss the tabloid press stories for their inaccuracy and inattention to detail, but the fact that these stories appear shows that there is a serious issue behind them. They underscore how serious the problem is, and they form a background to professional discussion of software reliability and how to achieve it. Here are a few samples gleaned from past issues to give a flavor of this discussion (volume and issue numbers are given in parentheses).[15]

Soon after the North American Air Defense System (NORAD) went on line in 1960, it detected a massive attack of incoming Soviet missiles and issued a warning to U.S. defense forces. The "missiles" were found to be the rising moon. A radar station in Thule, Greenland, picked it up as it rose over the horizon, but the NORAD computers had not been programmed to ignore it. (*SEN* 8/3).[16]

In November 1979 NORAD again signaled a phantom enemy attack and put the U.S. defense forces on alert. This time it was found that simulated data was introduced into the system to test its responsiveness, but due to a combination of equipment malfunctions and programming errors, the data was interpreted by other parts of the system as real (*SEN* 5/3).

The incorrect sign of a number representing latitude caused three jet fighters to flip over as they flew over the equator on autopilot (*SEN* 8/5). (This was referred to as an "old anecdote," and a subsequent column issued a disclaimer stating that the incident never really happened but that it *would* have happened had the error not been corrected in time.)

While a United Airlines 767 jet was descending for a landing, both its engines overheated and had to be shut down. The plane descended unpowered for four minutes until the pilot was able to restart them. The problem was later traced to a computer system that controlled the engine's power. The computer worked so well that the engines did not have to work as hard as they otherwise did, and so they did not generate enough heat to keep ice from building up on them. The accumulated ice blocked the flow of air and caused the engines to overheat (*SEN* 8/5). Note that the software in this instance worked perfectly and did exactly what it was supposed to do. The error was one of not thinking through the consequences of its working so well. Note also that when these kinds of things happen, the easiest way to fix them is by changing the software, not the hardware, and certainly not by telling the pilot to watch out for future occurrences!

Gemini V, which carried astronauts Gordon Cooper and Charles Conrad on a record-breaking flight in 1965 in preparation for a mission to the moon, splashed down a hundred miles from its intended target. The Gemini guidance equations used the sun as a reference point, and the programmers assumed that every twenty-four hours the earth would be at the same place relative to it. But this assumption neglects the earth's orbit around the sun, which makes for a difference of about one degree of longitude per day. Previous Gemini flights had been short and so the problem had not manifested itself, but it did on Gemini V, which lasted eight days (*SEN* 9/1).

The first launch of the Space Shuttle, scheduled for April 10, 1981 was scrubbed ten minutes before launch. It was later found that the primary and backup computers were out of synchronization by a few milliseconds and thus were unable to communicate with each other. This error had been introduced by a programming change made two years earlier, and the link between this change and the problem of synchronization went undetected, despite exhaustive testing and simulation of the Shuttle's computer system. Two days later, after the error was isolated and corrected, the *Columbia* blasted off the pad at Cape Canaveral on its way to a successful flight (*SEN* 6/5).

In September 1983 Korean Airlines Flight 007 was shot down by a Soviet interceptor after the Boeing 747 strayed over sensitive Soviet military installations. The reason the plane was off course may never be determined. But the most plausible explanation is that the crew made a mistake when keying coordinates into the plane's inertial-navigation computer. The mistake might have been a simple transposition of digits, like keying in 13245 instead of 12345. Then either because the pilots did not pay attention or because they deliberately shut off the alarm system, they did not respond to other warnings as the plane strayed off course (*SEN* 9/1).[17]

And finally, let me quote verbatim from *Software Engineering Notes:*

Monkey Business in 747 Macaquepit. *A monkey slipped out of its cage aboard a China Airlines 747 cargo plane before landing at Kennedy airport, and was at large in the cabin. After landing, the pilot and copilot remained in the cockpit until the animal control officer thought he had the monkey cornered at the rear of the plane. After the pilot and copilot left, the monkey then entered the cockpit and was captured while sitting on the instrument panel between the pilot and copilot seats. [The standard preflight checkout must also cover alterations made by inquisitive macaques!] (SEN 12/4 [Oct. 1987], p. 4)*

As the examples show, "Risks to the Public" mixes the serious and the whimsical, the fact and the folklore, in its selection of anecdotes. Beyond that, the column illustrates that computer errors, so familiar to the public, are not the same thing as software bugs. In many cases the program did exactly what it was supposed to do, but the system failed when it encountered conditions unforeseen by the designer. Others, like the hypothetical scenario of Korean Airlines Flight 007, were caused by human beings who inadvertently or even deliberately took steps to bypass the computer system's safeguards. In other instances the computer and its software were used as a convenient scapegoat to hide the real cause of failure, which might have been more embarrassing. In a discussion of the false alarms that have plagued NORAD from the time of its inception, one computer scientist wrote to the editor of "Risks to the Public":

All of these missile alerts were caused by real flying objects, hardware failures, or human error. I'm not saying that bugs didn't cause any missile alerts, just that the ones that are reputed to have been caused by bugs in fact were not.

Perhaps computer software—as opposed to the combination of hardware, software, and people—is more reliable than the folklore has it.[18]

The writer of those remarks is correct but is ducking the central issue. Computers by design are logic machines and tolerate little ambiguity in the way they interact with humans. That interaction is defined in the design of computer software. The procedures of setting switches, reading dials, moving levers, and typing commands into a keyboard cannot be divorced from the programming code that specifies the computer's actions, whether it be in BASIC, Fortran, or a more primitive machine language. In the words of one programmer, "There is no fundamental difference between programming a computer and using a computer."[19] Previous experience with other forms of technology, even those as complex as aircraft, has not been useful in defining the space where man and machine meet. *All* these questions of computer performance are rightly questions about software, for that is where they are resolved.

Setting the Stage

Current thinking about the human interface with computers bears the stamp of the early days of the U.S. manned space program: Projects Mercury, Gemini, and Apollo. How much a person will or can do, how much a person is aware of what the ship's automatic systems are doing, how much authority a person has to take over in the event of a malfunction or emergency—all these questions were asked and first answered in those programs.

Today it is hard to imagine how little was known about a human's ability to function in space when the United States embarked on its manned space programs after Sputnik. Animals, both American and Soviet, had flown in space, and in April 1961 Yuri Gagarin flew a complete orbit with no ill effects. But Americans knew few details of Gargarin's experiences, especially about the extent to which he had any control over the flight. Some of the concerns seem silly today; but in those days they had to be taken seriously.[20]

For Project Mercury, the first U.S. manned program, engineers designed the space capsule to be as automatic as possible. The astronaut just went along for the ride. Once injected into orbit by the Atlas booster, there was little anyone could do to alter the orbital parameters: the Mercury capsule had no orbital rockets and no on-board computer. Designers of Mercury did make provision for the astronaut to take over a modest control of reentry, but only if permitted to do so from the ground and after assurances that the person was alert and capable of performing the right actions. But such action by the astronaut was not necessary.

The folklore has it that the Mercury astronauts, most of whom were pilots accustomed to flying new aircraft past previous performance limits, rebelled against being treated like this, like "Spam in a can."[21] After the first few flights, however, it became clear to everyone that not only could a human being perform well in space, in some cases he *had* to in order to ensure a successful and productive mission. These flights set the tone of subsequent debates over the roles of man and computer when planning for the Apollo Lunar mission was going on. What emerged was a consensus to keep the pilot "in the loop" at all times, while letting the computer do as much work as it could. The current software interface with pilots and astronauts on the Space Shuttle and fly-by-wire aircraft is the child of those events.[22]

On February 20, 1962 John Glenn became the first American to orbit the earth. He suffered no ill effects of space travel whatsoever. Moreover, Glenn handled control of the yaw of the capsule all during the flight, and he brought it back to earth under manual control after ground controllers received an alarm that the capsule's heat shield was loose. To reenter the atmosphere, he had to fire the retro-rockets by setting switches and pressing a sequence of buttons on instructions from the ground. It was eventually discovered that the alarm was false. The first U.S. manned orbital mission provided two lessons. The false alarm showed that automatic safety systems are a mixed blessing. And Glenn's performance and safe return showed that there was nothing wrong with letting the astronaut take on some real flying if necessary.

On May 15 and 16, 1963 Gordon Cooper flew in the first U.S. long-duration flight into space. (Cooper would later fly in the Gemini V mission already described.) Like Glenn, Cooper also had to take over manual control of reentry, after a record-breaking twenty-one orbits. The reason was that a short circuit had cut all electrical power to the automatic stabilization system. Cooper executed a perfect reentry and back on the ground said, "Man is a pretty good backup system to all these automatic systems." A few days later President Kennedy presented Cooper with a medal, saying, "I think that one of the things which warmed us most during this flight was the realization that however extraordinary computers may be, we are still ahead of them, and that man is still the most extraordinary computer of all."[23]

Cooper's mission was the last Mercury flight. The success of Mercury led on to Project Gemini, an earth-orbital program designed to test rendezvous and docking techniques so critical to a mission to the moon. The Gemini capsule carried two astronauts, and a fifty-eight-pound transistorized, digital computer. And the Gemini and Apollo missions were planned with a capability for astronaut control right from the start.

Apollo

One final example illustrates how serious as well as how complex the issue of software reliability has become. It happened during the most dramatic moment in the history of space exploration, the first manned landing on the moon in July 1969. That landing came close to not happening, and the Lunar Module's computer and its programming played a role.

Programmers at the MIT Instrumentation Laboratory wrote the Guidance Computer software. Their first programs were simple utilities designed to test the computer and to drive the Display and Keyboard (DSKY); gradually the lab tackled the more complex jobs of deep-space navigation. Programmers knew each program by a colloquial name, most having some reference to the sun in honor of the Greek god Apollo: Sun-disk, Eclipse, Corona, Sundial, Aurora, and about a half-dozen others. The two main programs were Colossus, carried on the Command Module's computer to handle guidance and navigation, and Luminary, carried on the Lunar Module to handle the lunar landing and rendezvous.[24]

As with all Apollo hardware, each piece of program code was accompanied by a thorough review process, with signatures of approval required from several layers of management before releasing the code to the group that loaded it into the computer. Changes to a program, whether due to a change in mission objectives, the finding of an error during tests, or some other reason, were made only after formal submission and approval first of a Program Change Request then a Program Change Notice, which indicated why the change was needed, what impact it would have on other parts of the program, and what effect it would have on the computer's memory capacity and processing speeds. This system was in every respect identical to the Engineering Change Notices so common to the design of hardware for air and space craft and is clear evidence that the MIT programmers saw themselves as engineering the software just as others were engineering Apollo hardware.

The programs themselves were written in a language especially developed for the computer and were translated into binary code by a mainframe (an IBM 360) at the Lab. By present standards it was a primitive way to program: there were no standard high-level languages, and they used informal and colloquial names for variables and loose structures with many jumps in control. It took some firm pressure from NASA on MIT to pay heed to formal methods of program documentation and review. But from that experience emerged a system that maintained a level of software engineering few others could match in those days.

The binary code generated by the mainframe was then loaded into the computer's memory. This memory was called core rope: it stored the program by the pattern of wires woven around small doughnut-shaped cores of magnetic material. When the computer was being designed in the early 1960s,

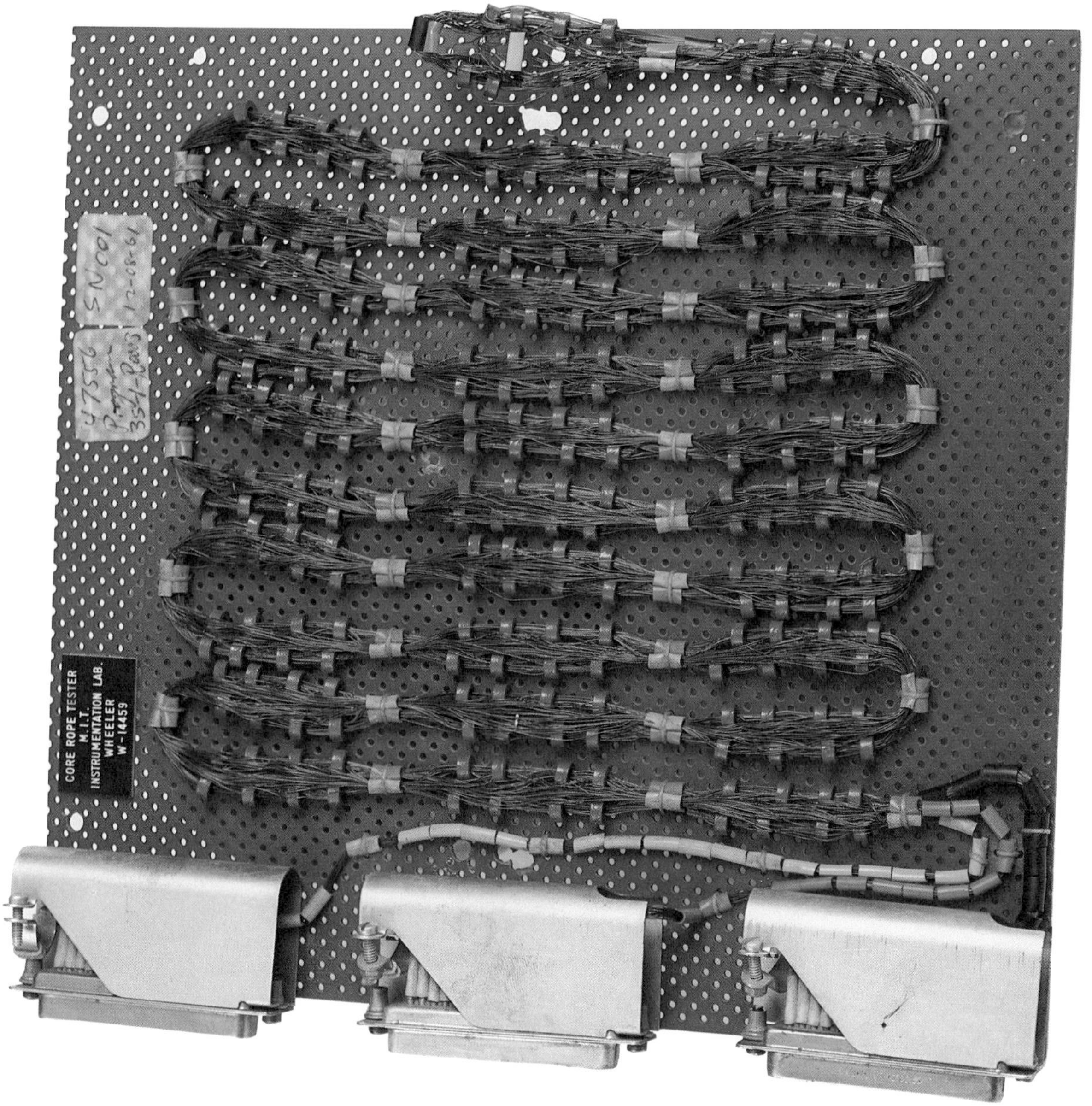
CORE ROPE TESTER
M.I.T.
INSTRUMENTATION LAB.
WHEELER
W-14459

9.1

The Apollo programs were literally woven into the computer as patterns on a sequence of memory cores. This is a prototype core rope tested at the MIT Instrumentation Laboratory in 1964. (Photo: David E. Dempster)

this method offered the best combination of reliability, compactness, and capacity for storage (figure 9.1). Because the pattern had to be physically woven into the ropes and then installed into the computer, the method put added pressure on the programmers to get things right. (In retrospect some have called this a healthy tonic that many other software projects lack.) Once the Apollo Command and Lunar Modules were stacked atop the Saturn booster in the huge Vehicle Assembly Building at Cape Kennedy, changing the program was extremely difficult if not impossible. Programmers at the Massachusetts Institute of Technology had up to six weeks before launch to make any changes and install new ropes. Yet they did provide themselves with a small escape: they designed the Apollo Guidance Computer so that data stored in the computer's small read-write memory, normally used for temporary storage of in-flight data, could be interpreted as program commands. (This provision came in handy on at least two missions: Apollo 13, when the Lunar Module computer had to take over the Command Module's navigation duties; and Apollo 14, when some last minute reprogramming of the Lunar Module's computer prevented a false alarm from automatically aborting the landing.)

The Apollo 11 Landing

Eldon Hall has called the Lunar Module Computer the first all-digital autopilot, the first system in which a real-time digital computer actively controlled a manned flying machine. This computer made critical decisions about the thrust and steering of the Lunar Module's thrusters and main engine during the descent onto the moon's surface. That was its primary mission, and that was why it was installed on the Lunar Module in the first place.

The computer had another function, that of assisting with rendezvous with the Command Module in lunar orbit after blasting off from the moon. Rendezvous and docking are no small task either. They were the reason for the on-board computer in the earlier Gemini spacecraft. During a landing there was no need for this function, so the preliminary mission plan and checklist had the astronauts flip a switch to disable it just before beginning the descent.

If the landing had to be aborted for any reason, the computer would have had to switch over to the rendezvous program quickly. For this reason mission planners decided fairly late in mission planning to leave the rendezvous radar in active mode, and they changed the checklist accordingly.[25] The Lunar Module computer was capable of locking on to the Command Module without the radar being active, but this change added just a bit more safety.[26]

Although the landing had been rehearsed again and again in simulators on the ground, these tests failed to reveal that the combined data from both the rendezvous radar and the landing would overload the computer. When the Lunar Module fired its rockets to begin the final descent, signals from the radar added thirteen percent more work on the computer than its programmers had anticipated. According to plan the computer shed some of the less-critical tasks and devoted its processing primarily to the landing. On the Lunar Module's Display and Keyboard (DSKY) and on screens at Mission Control in Houston there flashed "1201–1202 alarm," which indicated that there was an overload and that the computer was restarting its processing of the most critical tasks before it, the landing (figure 9.2). Again, this response by the computer was what its programmers had designed it to do in case of an overload. The alarm did not, however, tell the astronauts or the ground controllers why it was occurring, nor did the alarms indicate how close the computer was to being so overloaded as to affect the landing program.

In fact, the computer was handling the descent safely, though no one was sure of that. Edwin Aldrin, the Lunar Module pilot, called out these alarms to Mission Commander Neil Armstrong as the computer flashed them on the display. The two men checked to see what the alarm meant, and conferred with Steve Bales, who was monitoring the alarms at the Manned Spacecraft Center in Houston. Bales monitored the same alarms that the astronauts saw; he could also monitor a lot more about the computer's performance, though there was a six-second delay between the Lunar Module and his console

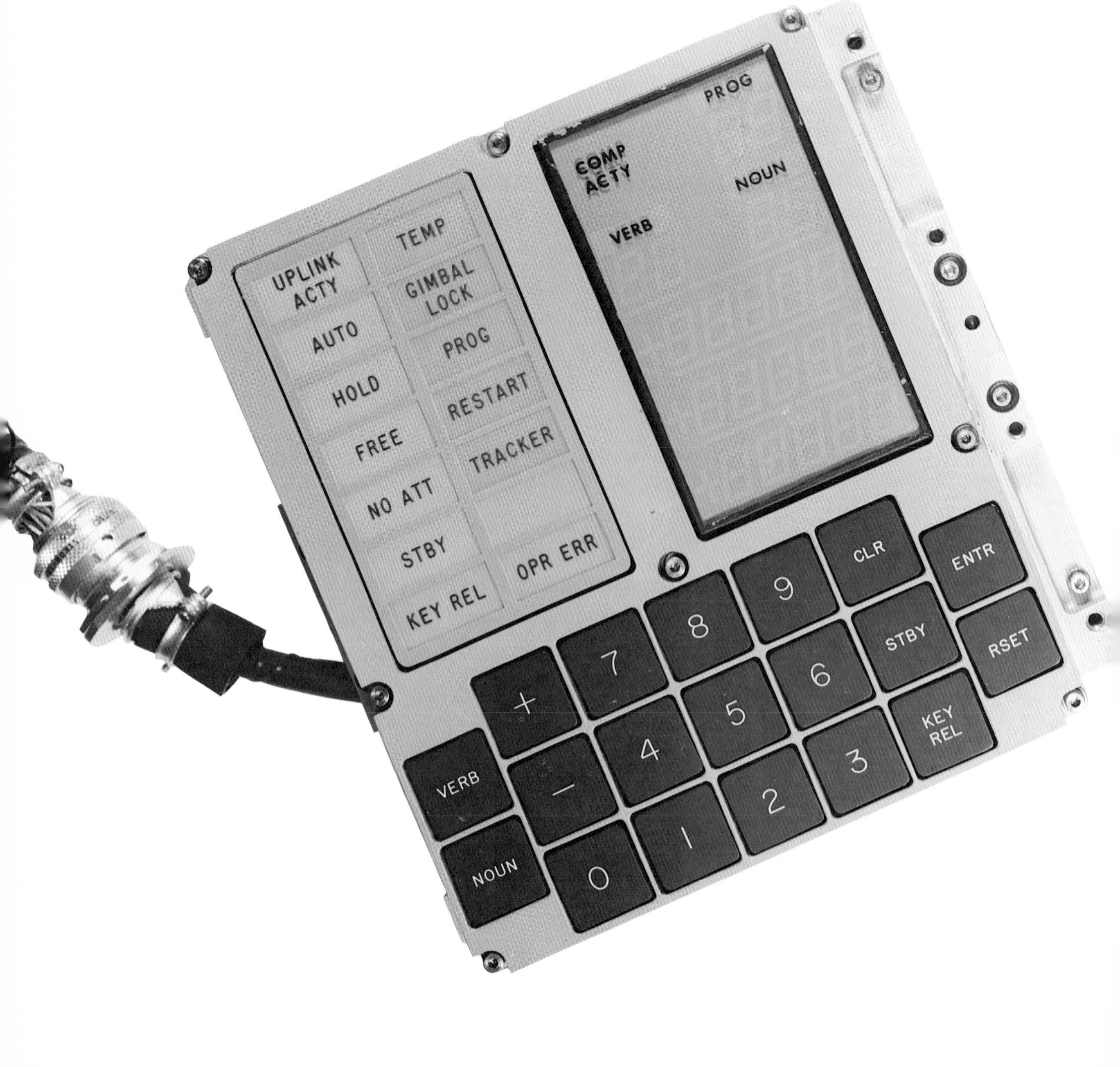
PROG
COMP ACTY
NOUN
VERB
UPLINK ACTY
TEMP
GIMBAL LOCK
AUTO
PROG
HOLD
RESTART
FREE
TRACKER
NO ATT
STBY
OPR ERR
KEY REL
CLR
ENTR
9
8
7
+
STBY
RSET
6
5
4
KEY REL
3
VERB
−
2
1
NOUN
0

9.2

Apollo astronauts communicated with their on-board computer via this DSKY (Display and Keyboard, pronounced "diskey"). Commands were given numerical codes and were entered in as noun-verb combinations (e.g., "display fuel remaining"). (Photo: Charles Stark Draper Laboratory)

in Houston (figure 9.3). Yet he did not have enough information to assure him that the landing was proceeding safely. He and his fellow controllers did notice that all the other parameters of the descent were within safe limits, and that the 1201 and 1202 alarms were only intermittent and at "relatively" infrequent intervals (in all, the alarms flashed less than a half-dozen times during the descent). Mission controllers also knew that the Lunar Module's Abort Guidance System was working and that it could safely return the two men to lunar orbit at any time. This backup computer could not have handled a landing, but its presence did mean that the problems with the main computer were not irrevocably putting the astronauts' lives at risk.

Bales recommended that Armstrong proceed with the landing. But precious minutes had been lost, and the astronauts had been distracted from the main business at hand of bringing the Lunar Module down to the surface. When Neil Armstrong finally got a chance to look at the programmed landing site (chosen from photographs), he found it strewn with boulders. He then took over manual control of the craft, guided it over the boulders to a smooth spot, and landed safely. He had twenty-four seconds' worth of fuel left.[27]

The overflow alarms nearly cost the whole mission, and it occurred precisely at the most critical moment of the mission. What happened? Clearly the rendezvous radar caused the problem (on future missions the astronauts switched it off before the final descent). But the astronauts should never have been dragged into a problem with rendezvous during the landing. Back in Houston, Steve Bales had to decide that the alarm was not going to jeopardize a landing. Could not the computer have made that judgment and thus avoid distracting Armstrong and Aldrin? On the other hand, the astronauts had asked to be given as much information and control from the computer as they could handle. Armstrong did land safely after he, with advice from other humans, ignored the warning and landed manually.

This story, with its implied criticism of Apollo software, has been circulated among computer professionals as another example of how difficult it is to know that a program will work as

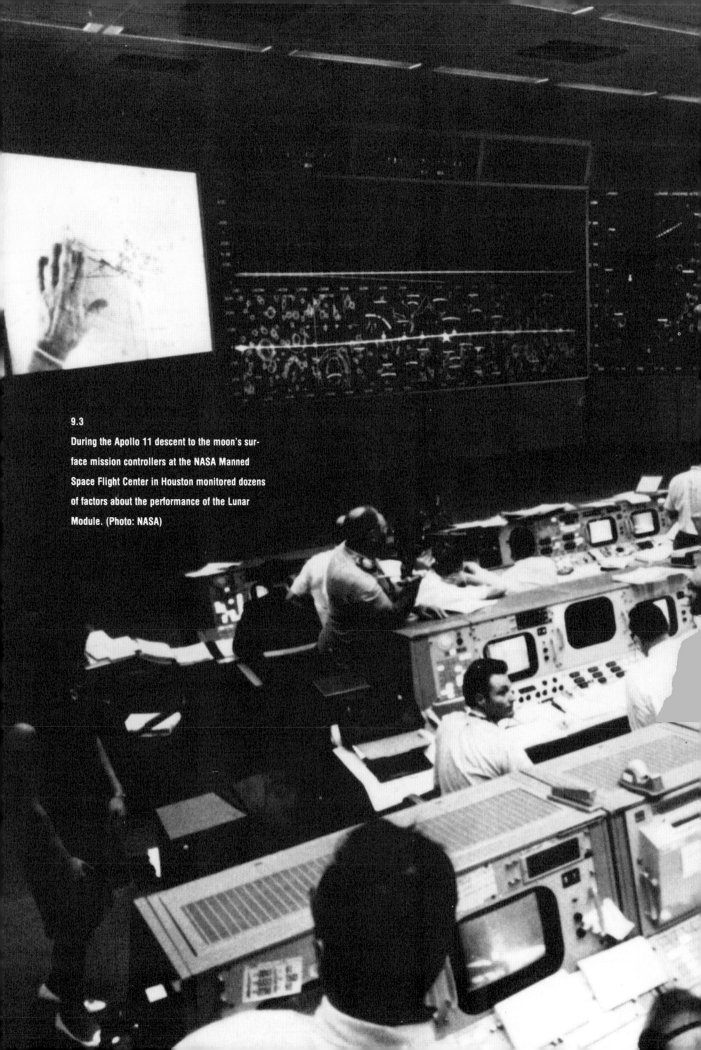

9.3
During the Apollo 11 descent to the moon's sur-
face mission controllers at the NASA Manned
Space Flight Center in Houston monitored dozens
of factors about the performance of the Lunar
Module. (Photo: NASA)

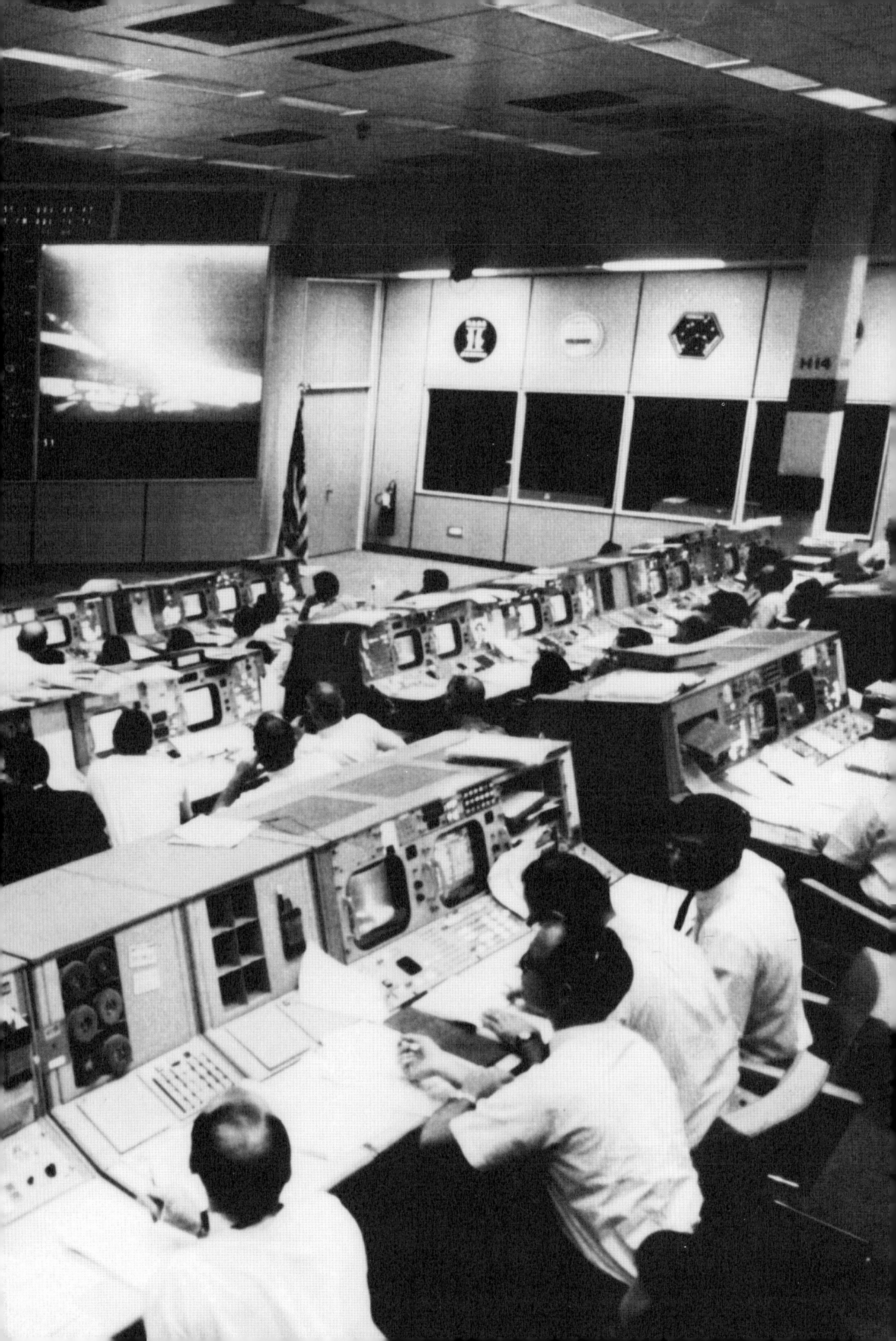
H14

anticipated. In Software and Its Impact in the trade journal *Datamation*, Barry Boehm cited it as an example of how "some of the most thoroughly tested" software can still cause problems.[28] A few months later Margaret Hamilton, a member of the Apollo programming team at Draper Labs, replied, "The APOLLO 11 problem referred to by Dr. Boehm was not a software problem. In fact, it was the software that prevented a potentially dangerous situation." Here she is referring to the fact that the programmers had provided for an automatic restart in the software, which allowed the computer to shed the less-critical tasks to concentrate on the landing.

Following her letter to the editor was Boehm's reply, which said in part:

The APOLLO 11 situation surfaces one of those fundamental questions about the nature of software: "Is an error of omission in system design, that is not caught during software development and testing, a software error?" You can resolve it either way; however consistency would suggest that to the extent that software deserves to take credit for "preventing a disastrous situation" on APOLLO 11, the software also deserves the responsibility for diverting Armstrong during the Lunar landing phase.[29]

Current debates over the reliability of software for the Strategic Defense Initiative have the same tone as the preceding exchange between Margaret Hamilton and Barry Boehm. A lot of progress has been made in software engineering in the intervening years, but the basic nature of the issue is as contentious as ever.[30] Perhaps the issue is not so much whether software is inherently reliable but that human beings are by nature unused to the need to be so precise. The computer's intolerance of ambiguity made us aware only of something that was always there but hitherto hidden from our consciousness. (It appears in folktales like the Sorcerer's Apprentice, for example, in which a literal obedience to the boy's commands gets him into trouble.) The real lesson of the Apollo 11 landing was not that reliable software cannot be written; it was that the variety and richness of actual experience in flight or elsewhere can never be fully captured in a simulation or other model.

Aerospace engineers have had to accept the fact that the software they work with is not perfect. But as we have seen in our look at aerospace design, that is not as foreign to that community as it might be to other engineering disciplines. The structural engineer has the impossible task of making an aircraft strong yet lightweight; he does not have the luxury of respecting one demand while ignoring the other. Aerospace software seems to share that quality. The aerospace programmer has the impossible task of writing programs that are reliable yet run in real time. That aerospace engineers continue to fly systems in the face of this impossible task horrifies computer professionals on the outside.[31] But that will not stop others from going ahead. Perhaps the best one can say about aerospace software is that in addition to requiring all those things that ordinary software requires, it also requires a measure of the same kind of bravado that causes one to want to fly in the first place.

Chapter 10
Conclusion

The computer and the aerospace industries have each driven the other at a furious pace since the end of the Second World War. The relationship has been mutually beneficial. But on the whole I believe that the computer industry has benefited more. Projects like Whirlwind and SAGE gave the computer industry the capability of designing and manufacturing computer systems on a large scale. Also, the programming requirements for those systems laid the foundation for the industrial production of computer software. Many current large-scale programming methods and techniques are derived from Air Force and NASA projects. Apollo is only the most visible.[1] The computer industry owes its tremendous growth in more recent years to the invention of the integrated circuit, again a development that received Air Force and NASA help at its most critical moment.

The computer industry owes a debt to the U.S. Air Force and to NASA, who both sponsored expensive and risky projects at crucial moments, projects that had enormous returns not just for themselves but for the rest of society as well. But in recent years this phenomenon has diminished. Some current large-scale aerospace computing projects, like the Air Force's Very High Speed Integrated Circuit program and the Strategic Defense Initiative, do have a potential for raising the level of U.S. computer expertise. No one is sure how to make that happen. But in any case, the spin-offs are not the primary goal of the projects. In short, the computer industry no longer needs aerospace to survive. Ironically, this has been the case since about 1975, just as the Apollo program came to an end.[2]

Other current Defense Department initiatives have been either resisted by the computer industry or else reluctantly accepted only because of the money they bring with them. The computer language Ada (r) is a case in point.[3] In 1983 the Department of Defense formally designated this language as one it would require all future military systems to use. They did so to remedy the inefficiency and expense of maintaining up to four hundred separate computer languages throughout military computer systems. Ada is now being used, but it has yet to demonstrate its advantages over other languages. The entire program has been criticized by computer scientists as unworkable.[4]

As the computer industry has matured, it has been able to raise the capital and manpower to push ahead with its own agenda for technical innovation. Computer companies like Tandem, for example, have developed fault-tolerant machines, which in many ways are superior to the multiple, redundant processors that NASA adopted for the shuttle a decade ago. These companies have not needed a direct push from aerospace to get on with this work.

Benefits to Aerospace

Computers have enabled the aerospace industry to do many things it had always wanted to do but could not in an earlier day. Going to the moon was the first. Powerful CAD/CAM systems are now at the heart of all current aerospace engineering. Fly-by-wire controls have greatly extended every corner of the performance envelope of both air and space craft. Indeed, without it a machine like the Space Shuttle could not be flown.

But introducing the computer into the matrix has not made the aerospace business any simpler. It is harder than ever to make money building and selling airplanes, and the U.S. space program has fallen on equally hard times since the last Apollo mission. As of this writing, the highest-flying, fastest airplane in the United States is still the SR-71, designed over twenty years ago by Clarence "Kelly" Johnson and his team of engineers at Lockheed's "skunk works" outside Burbank. For that and for all the other famous airplanes he designed, Johnson's primary computational tool was a ten-inch slide rule (figures 10.1 and 10.2).

CAD/CAM makes it much easier to incorporate engineering changes in a design, but in doing so it requires someone on a design team to know when to freeze the requirements and resist the temptation to add just one more feature, which inevitably adds weight and ultimately compromises the whole project. Some recent machines, like the C-5A Transport, the B-1B long-range bomber, and the proposed U.S. Space Station, have suffered from a lack of such discipline. Other recent

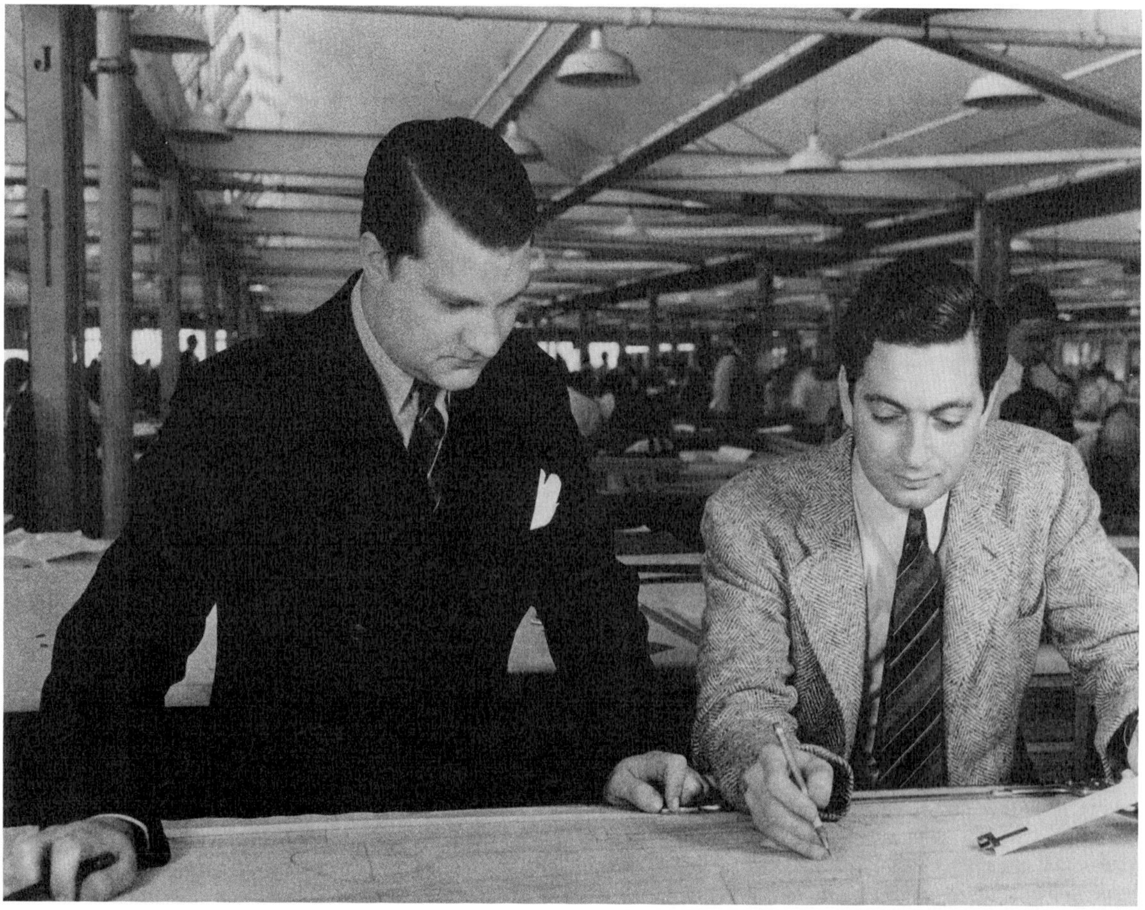
J

10.1

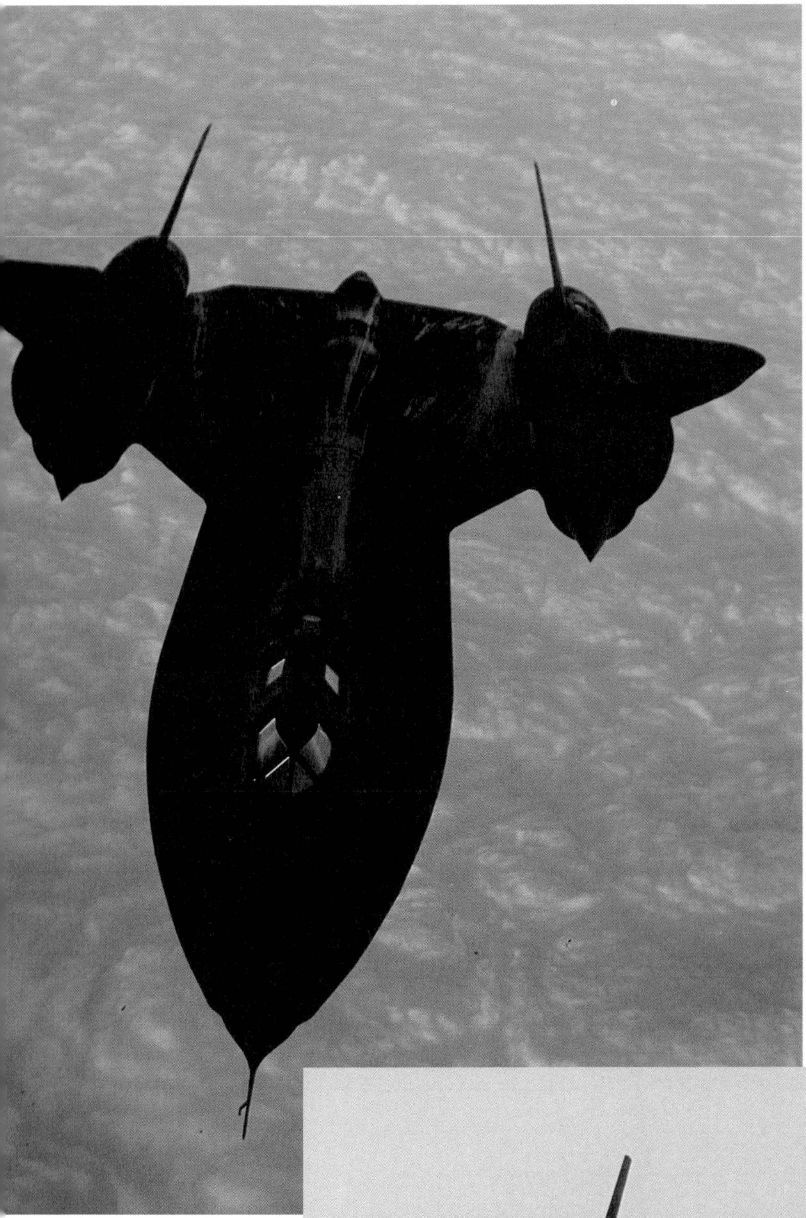

10.1

Clarence "Kelly" Johnson, who designed aircraft at Lockheed from 1933 until his retirement in 1975, was responsible for a host of famous aircraft designs, including the triple-tail Constellation, the F-104, the SR-71 Blackbird, and the U-2, just to name a few. Here he and Phil Colman (holding the slide rule) are going over the design of the P-38, one of the most successful Allied fighter planes of the Second World War. (Photo: Lockheed Corporation)

10.2

The SR-71 Blackbird, a high-altitude reconnaissance airplane designed by Kelly Johnson in 1961, remains the highest-flying, fastest aircraft in the world. (Photos: U.S. Air Force)

10.2

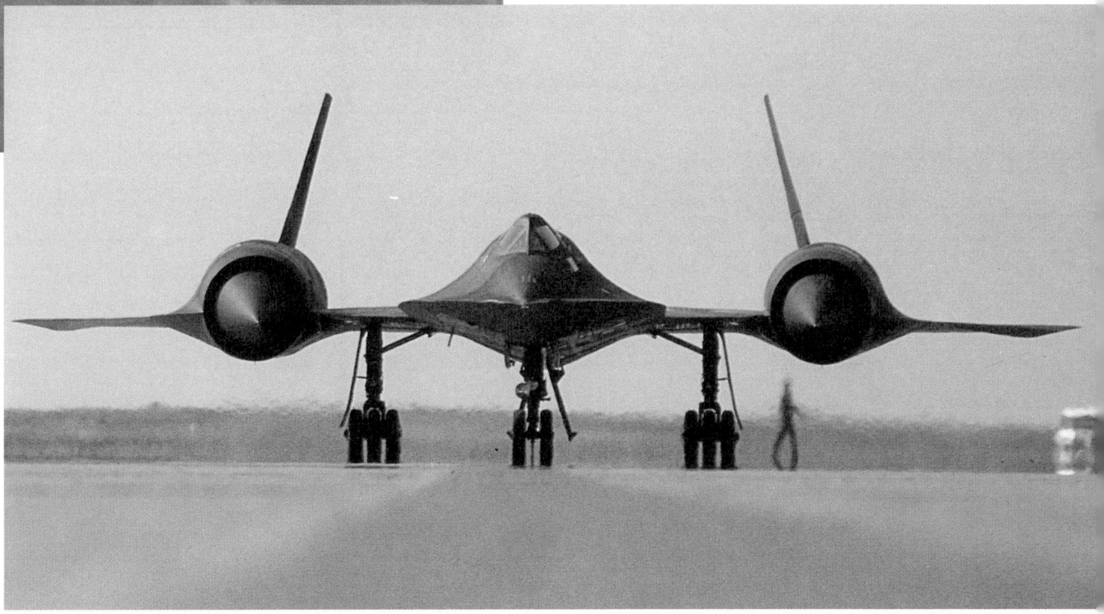

programs, like Boeing's 757 passenger jet, have been commercial and engineering successes, and the computer shares much of the credit for those successes. CAD/CAM has made it easier to fail as well as to triumph on a grand scale.

Computers, it seems, tempt aerospace engineers to violate one of the most sacred rules of sound engineering, that is, to keep things as simple as possible while satisfying all the conflicting demands of a design specification. A simple design is usually more reliable and can be manufactured and maintained at lower cost. A simple design is also often a lightweight design, which immediately confers performance and cost benefits. Computers by nature are machines that handle the complex. That was why they were invented, and that is why we depend on them so much. As a design tool, computers are best used when they help a designer sort through a maze of details to come up with a workable design that is also simple. But there is no guarantee that using a computer will automatically lead to a simple design. To keep things simple and lightweight in an environment of powerful CAD/CAM tools requires single-mindedness and discipline that not every designer or management team has.

The public encounters aerospace and computing technology at the airport more than anywhere else. The daily press is filled with stories about the discomforts of modern commercial air travel and of the woes that plague the air traffic control system. Usually the blame is laid on the recent deregulation of the industry, with a dose of stories about computer snafus thrown in. The deregulated airline business, with its numerous discount fares and its hub-and-spoke scheduling arrangement, requires the assistance of some of the most powerful computer systems in the world, both to book the seats and to handle air traffic control. These computer systems work well, although there is much room for improvement. Airline passengers feel inconvenienced, yet they fill the planes. Those who complain the most are probably those who recall the days when planes flew mostly empty. Deregulation has opened up air travel to many people who could not afford to fly at more expensive fares. In all of this, the computer makes deregulation possible.

The application of inertial-guidance techniques to air and space craft is the best example of a direct impact of computer and microelectronics technology on aerospace. The rapid perfection of this technique took spacecraft designers and policy-makers by surprise. Inertial guidance, by being accurate, immune to interference, small in size, rugged, and relatively cheap compared to competing systems, forced its way onto missiles and now commercial airplanes.[5] The results have been widespread and diverse: from the reduction in crew size for long-range passenger planes to a destabilization of the nuclear balance of terror between the United States and the Soviet Union. In nuclear weaponry the computer has had a major impact, but the record shows that the people involved in these projects were not always in full control of the pace of innovation.

Beyond the Limits

It was not just the sound barrier that pilots broke through in the decades that followed the Second World War; it was a host of limits that aeronautic and space engineers were able to penetrate. And in each instance the digital computer was there. Its effect has been not just to break through a prewar limit of performance but also to raise the state of the art to a new plateau, from which designers mount assaults on new limits scarcely imagined before. That process is still going on. It has not been a story of unbroken and benevolent progress. And it is by no means certain that the next two decades will bring progress in flight at anything like the pace already set. But if people choose to pursue aims like permanent colonies in space or on the moon, a manned mission to Mars or one of the other outer planets, a new generation of cheap, safe, and fuel-efficient private airplanes, just to name a few, the computer will be there (figure 10.3).

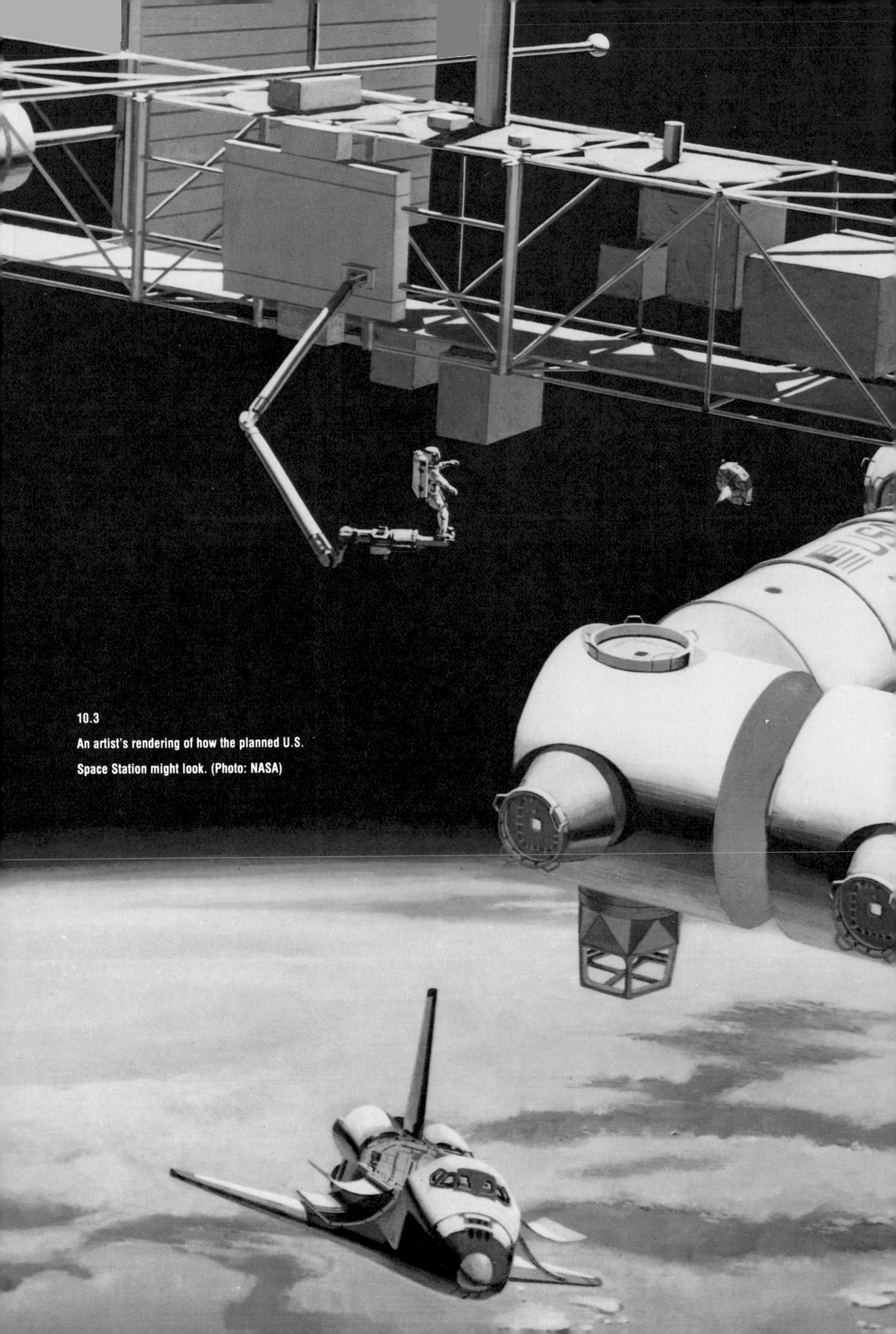

10.3

An artist's rendering of how the planned U.S. Space Station might look. (Photo: NASA)

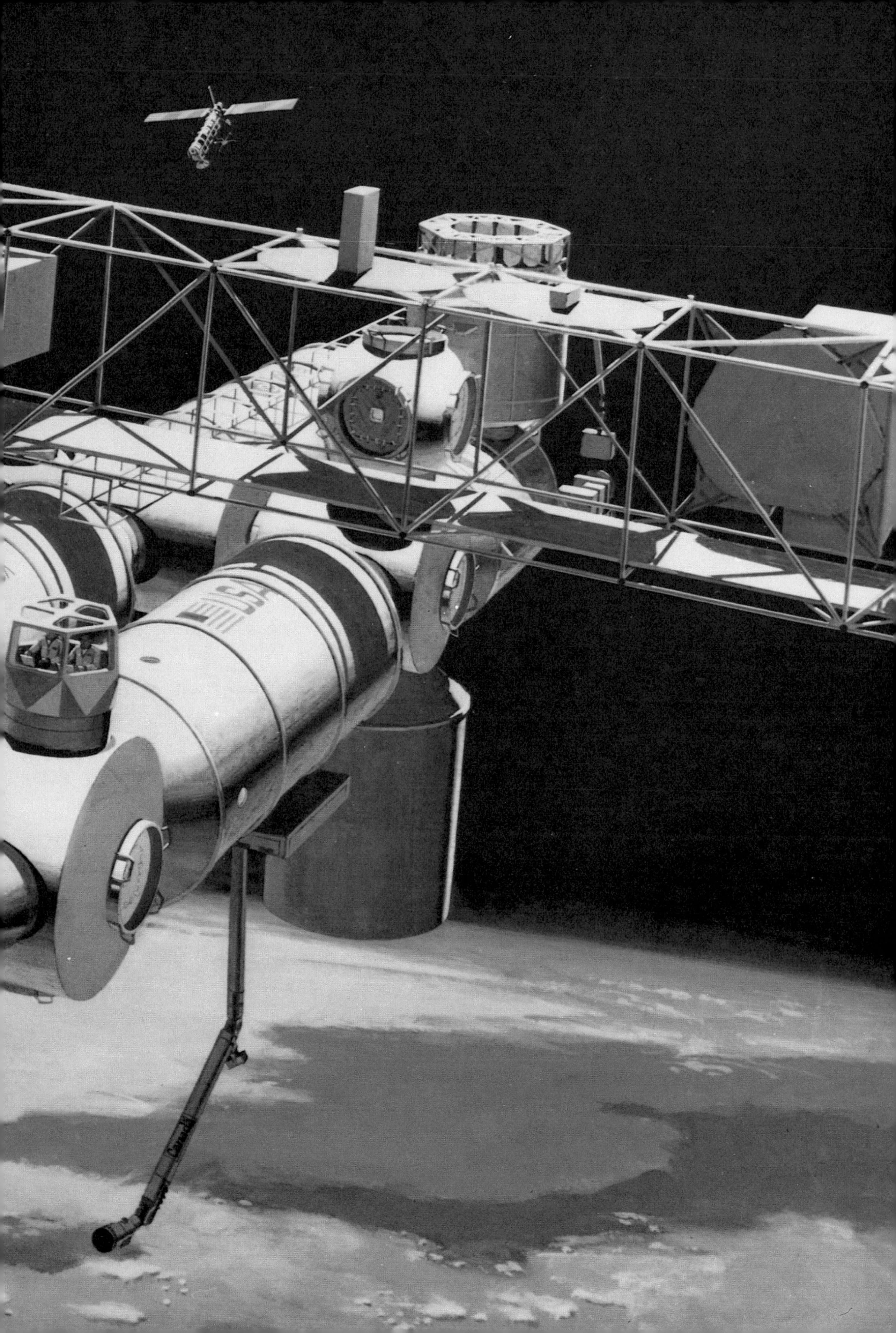
USA

Appendix

The Role of Computers in the *Challenger* Explosion

On January 28, 1986, the Space Shuttle *Challenger* exploded just over a minute into launch, and all seven crew members on board were killed. After an exhaustive investigation the cause was traced to hot gases burning through a seal in one of the solid rocket boosters. The escaping flame burned through the external hydrogen tank, and eight seconds later the entire solid booster broke away from its attachment to the external tank. The resulting structural failure led to the explosion of the remaining liquid hydrogen and oxygen and to the destruction of the Shuttle Orbiter itself.[1]

Motion pictures taken of the launch showed evidence of escaping gases within seconds of launch. Yet all through the launch ground controllers monitoring the Shuttle's computers noticed nothing unusual right up to the moment of the explosion. Telemetry did reveal that the thrust in one of the solid boosters had dropped. During the ascent the Shuttle's onboard computers compensated for this imbalance by swiveling the nozzles of both main engines and of the two solid boosters. This deviation was within the limits of the Shuttle's performance abilities, and so it did not cause an alarm to flash on any of the controllers' screens or send an alert to the astronauts.

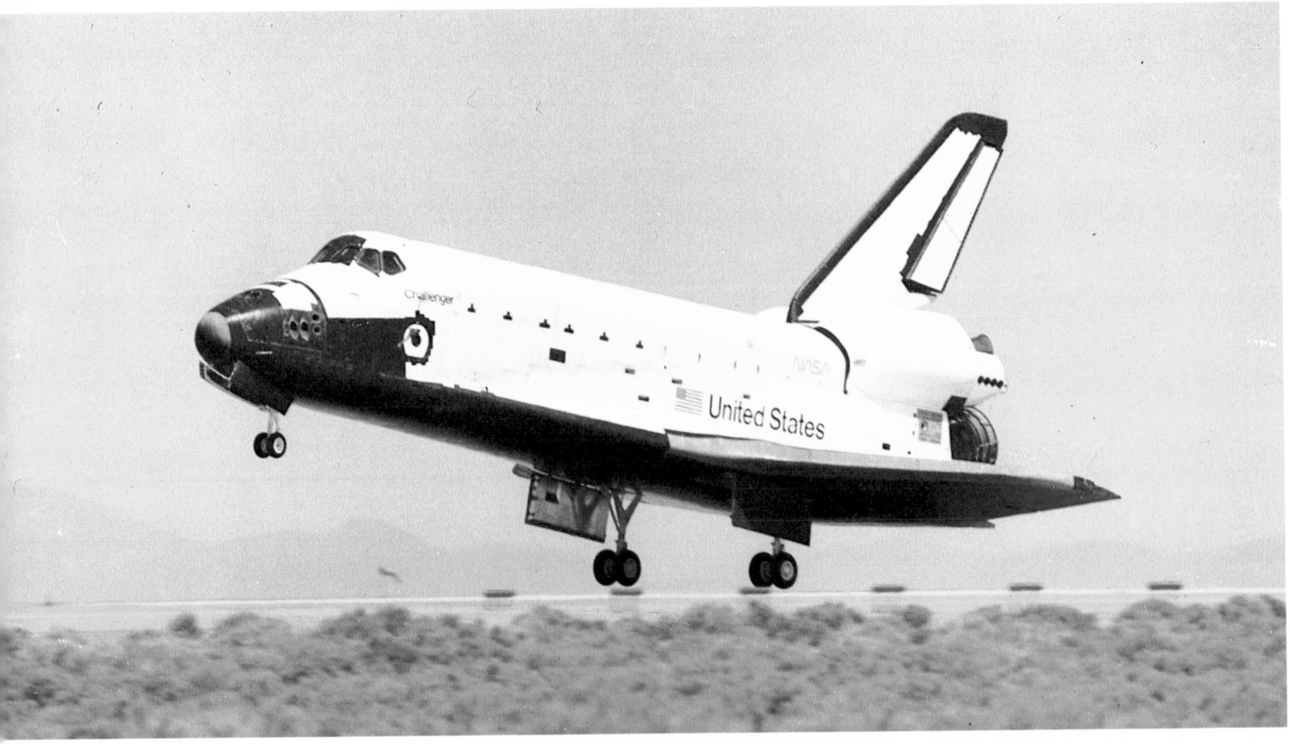
United States

a.1

The Space Shuttle *Challenger*, making its first landing on April 9, 1983 at Edwards Air Force Base, California. (Photo: NASA)

Could the computers have detected this problem and done something about it to save the crew? The simple answer is no. Even if sensors had picked up the escaping gases, it would have been impossible safely to abort the mission. By design, the Shuttle's solid boosters cannot be shut down once ignited; they have to burn until their fuel is spent. Again by design, there is no provision for separating the orbiter from the boosters in the event of a problem during ascent. Even if the crew could have separated the orbiter before the explosion, it is unlikely that the craft would have survived the resulting aerodynamic stress as it broke free while traveling at nearly twice the speed of sound. And if it did survive, the chances were remote that the crew could have maneuvered it to a safe landing back on the ground (or more likely, in the ocean).

Previous experience with manned space launches has led to a policy of not letting the reading from a sensor automatically abort a mission once begun. Before a launch, indeed right up to the milliseconds before liftoff, the computers are given authority to halt the countdown automatically. In the past this authority has prevented disaster by shutting down a bad engine seconds before it might have exploded. The computer acted much faster than any human controller could. But once the rocket has lifted off the pad, engineers are reluctant to let automatic systems retain that authority. The computer systems of the *Challenger* thus lacked this control (figure a.1).

The reason is that sensors and computers can fail and give false readings, especially during the mechanical stress of a launch. When a false alarm halts an otherwise smooth countdown, as has happened, the result is frustration and frazzled nerves but no unnecessary risk of human life. But during a launch an abort sequence is very risky. As noted in previous chapters, such false alarms occurred throughout the Apollo, Gemini, and Mercury missions, and in those cases the presence of a human being to detect and then override the false alarm was crucial to a successful mission.[2]

The *Challenger* was outfitted with a battery of sensors to monitor engine performance and detect problems if they occur. But few of these sensors were attached to the solid rockets, as the solid boosters were felt to be less of a problem than the more complex main engines. During early flights there were many more sensors attached to the solid boosters, but these were gradually removed as NASA gained flight experience. Remember that sensors increase the complexity, weight, and risk of a mission, and they must be employed sparingly to be effective.

The presidential commission found that the problem with the booster joints had surfaced on earlier shuttle flights, though not with catastrophic results. If sensors had been left on those boosters, they would have detected the problems and alerted engineers to fix them before the *Challenger* accident. But that would not necessarily have prevented the disaster. The commission noted that postflight examination of spent boosters from previous missions gave ample evidence of the problem. But corrective action was not taken in time to prevent the tragedy. Had telemetry and computer data been added to that evidence, perhaps it might have tipped the scales in favor of more caution or accelerated existing plans to redesign the booster joints. It seems reasonable to conclude that the lack of sensors and telemetry on the solid boosters was not a factor in the *Challenger* disaster.

The president's commission that investigated the *Challenger* noted these facts and concluded that the fault lay with NASA's decision process, which permitted the launch to go ahead. One of the commission members, Richard Feynman, wrote an appendix to the report, giving his personal observations on the reliability of the Shuttle. His conclusions were that the procedure for reviewing and validating the reliability of the rockets was faulty. In that essay he also described the review process for the development and validation of the Shuttle's software. Noting that software errors had cropped up, he writes:

The software is checked very carefully in a bottom-up fashion. . . . A discovery of an error during the verification testing is considered very serious, and its origin studied very carefully to avoid such mistakes in the future. Such unexpected errors have been found only about six times in all the programming and program changing (for new or altered payloads) that has been done. . . . To summarize then, the computer software checking system and attitude is of high-

est quality. There appears to be no process of gradually fooling oneself while degrading standards so characteristic of the Solid Rocket Booster or Space Shuttle Main Engine safety systems.[3]

One of the goals of the Space Shuttle program was to provide a rapid turnaround, so that an orbiter could return to space within weeks after a mission. That was a major change from the Apollo missions, which required months to check out the booster stages, Command, Lunar, and Service Modules after they were stacked atop one another at the Kennedy Space Center. The sheer size and complexity of Apollo demanded some degree of automation of the countdown. Compared to Mercury and Gemini, Apollo did incorporate quite a bit of automation. But an Apollo launch was still labor-intensive and time-consuming.

Advances in computers enabled Shuttle managers to reduce both time and manpower needs, though up to the *Challenger* explosion they never achieved the routine launch schedule they had hoped for. (In fact, the Shuttle program has never matched the record of Project Gemini, which required precisely timed, multiple launches of two-man capsules atop liquid-fueled Titan boosters.) After *Challenger* the whole question of a safe launch schedule was reexamined. The expectation that the system could be placed into routine operation like a passenger airline was premature. It is clear, however, that as designed, the Shuttle's on-board and ground-based computer systems performed well, as well as many other of the Shuttle's systems.

Notes

Chapter 1

1   Alex Roland, *Model Research: The National Advisory Committee for Aeronautics, 1915–1958* (Washington, D.C.: NASA, 1985), vol. 1, pp. 187–207; Frederick Ordway III and Mitchell R. Sharpe, *The Rocket Team: From the V-2 to the Saturn Moon Rocket* (Cambridge: MIT Press, 1979); Walter A. McDougall, *The Heavens and the Earth: A Political History of the Space Age* (New York: Basic Books, 1985), pp. 89–97, 166.

2   Alan Bromley, Charles Babbage's Analytical Engine, 1838, *Annals of the History of Computing,* 4/3 (July 1982), pp. 196–217. See especially p. 204, where Bromley concludes, ''The Analytical Engine could have been built with the technology at Babbage's disposal, although the work would undoubtably have been demanding and expensive.'' The opinion that Babbage's design was too far advanced for the existing technology has been widely stated; one of the most influential and earliest of these statements is by Howard Aiken, in the Introduction to *A Manual of Operation for the Automatic Sequence Controlled Calculator* (Cambridge: MIT Press, 1985), p. 7.

Chapter 2

1   E. T. Wooldridge, *Winged Wonders: The Story of the Flying Wings* (Washington, D.C.: Smithsonian Institution Press, 1983).

2   At high altitudes with proper shielding from the sun's glare it is possible to track a star in daylight. A star tracker can lock onto a specific star and keep it in its sights, using the same techniques that allow an antiaircraft gun to automatically track a target.

3   Letter from R. Rawlins of Northrop to the Electronic Control Corporation (Mauchly's company), June 25, 1947; quoted by Nancy Stern in *From ENIAC to UNIVAC: An Appraisal of the Eckert-Mauchly Computers* (Bedford, Mass.: Digital Press, 1981), pp. 117–118.

4   Richard E. Sprague, A Western View of Computer History, *Communications of the ACM,* 15 (July 1972), pp. 686–692.

5   Northrop Aircraft, Northrop Developments in Automatic Celestial Navigation, U.S. Air Force, Project MX-775, Report no. GM-536, Feb. 1, 1950; and an interview with Harold Sarkassian, Smithsonian Computer History Project, Sept. 11, 1972.

6   Northrop Aircraft, Quarterly Progress Report and Project Summary, April, May, June 1958, Snark Report no. GM-958.

7   Charles Babbage Institute for the History of Information Processing, *CBI Newsletter,* 7/1 (fall 1984), p. 6.

8   Richard E. Sprague, A Western View of Computer History, p. 687.

9   Interview with Paul King, February 1973, Smithsonian Computer History Project. Parameter variation, which formed the basis for much aerospace engineering, is discussed in Walter Vincenti, The Air Propeller Tests of W. F. Durand and E. P. Lesley: A Case Study in Technological Methodology, *Tech. Cult,* 20/4 (1979), pp. 712–751.

10   John W. Sheldon and Liston Tatum, The IBM Card-Programmed Electronic Calculator, in Brian Randell, ed., *The Origin of Digital Computers: Selected Papers,* 2nd ed. (New York: Springer-Verlag, 1975), pp. 229–235. Also Michael R. Williams, *A History of Computing Technology* (Englewood Cliffs, N.J.: Prentice-Hall, 1985), pp. 254–257.

Chapter 3

1   Although a computer is necessary to solve all but the simplest linear programming problems, the term "linear programming" should not be confused with computer programming, which is a very different activity.

2   George Dantzig, Linear Programming, a paper presented at the Symposium on Numerical Methods, Los Angeles, July 1948. See also Robert Dorfman, The Discovery of Linear Programming, *Annals of the History of Computing,* 6 (July 1984), pp. 283–295; Mina Rees, The Computing Program of the Office of Naval Research, 1946–1953, *Annals of the History of Computing* 4 (Apr. 1982), pp. 102–120.

3   George Dantzig, The Need for High Speed Electronic Computers for Programming, a proposal for a meeting of the Ad-Hoc Panel on Digital Computers, August 1949. National Air and Space Museum, National Bureau of Standards Archives 15-7.

4   Nancy Stern, *From ENIAC to UNIVAC* (Bedford, Mass.: Digital Press, 1981), pp. 107–108; National Bureau of Standards, *Computer Development (SEAC and DYSEAC)* (National Bureau of Standards Circular no. 551, Washington, D.C., 1955).

5   BINAC never performed more than a few test calculations, although it was delivered a year before SEAC. In mid-1949 Maurice Wilkes of the University of Cambridge, England, completed EDSAC, which is widely regarded as the world's first operational stored-program electronic computer.

6   L. R. Johnson, Installation of a Large Electronic Computer, *Proceedings ACM Meeting,* Toronto, Sept. 8–10, 1952, pp. 77–81. UNIVAC Serial no. 1 is usually regarded as the first computer delivered to a customer in the United States, but it remained at the Philadelphia factory until December 1952, when it was shipped to the Census Bureau in Washington. Parts of it are now at the Smithsonian Institution.

7   James L. McPherson, Census Experience Operating a UNIVAC System, in *Symposium on Managerial Aspects of Digital Computer Installations* (Washington, D.C.: U.S. Office of Naval Research, 1953), pp. 30–36. There was the further complication

that at this time Mauchly was being investigated by the FBI for alleged ties to the Communist Party. The charges were eventually found to be entirely without foundation, but in the political climate of the early 1950s even this suspicion was enough to prevent Mauchly's access to many of the potential customers of his own creation.

8   Fred Kaplan, *The Wizards of Armageddon* (New York, Simon and Schuster, 1983), pp. 58–64.

9   Material for this section has been compiled mainly from interviews conducted for the Smithsonian Computer History Project. See also F. J. Gruenberger, The History of the Johnniac, *Annals of the History of Computing,* 1/1 (1979), pp. 49–64.

10   See, for example, Allan Newell and Herbert Simon, The Simulation of Human Thought, RAND Corporation Research memorandum RM-2506, Dec. 28, 1959. There were many discussions in the 1950s of how (or whether it was possible) to get a computer to "think" like a human being. Newell and Simon were among the leaders in a technique of artificial intelligence that has since become common.

11   C. C. Hurd, Edited Testimony, *Annals of the History of Computing,* 3/2 (Apr. 1981), pp. 163–182; also Martin Weik, A Survey of Domestic Electronic Digital Computing Systems, U.S. Army, Ballistic Research Laboratory, Aberdeen, Md., Report no. 971, Dec. 1955, pp. 67–80.

12   Paul Armer, SHARE—A Eulogy to Cooperative Effort, *Annals of the History of Computing,* 2/2 (1980), pp. 122–129.

Chapter 4

1   U.S. Office of Naval Research, Special Devices Center, Project Hurricane, Quarterly Report no. 1, Oct. 1, 1948.

Chapter 5

1   Bernard Loveman, Reliability of a Large REAC Installation, *Proceedings, Eastern Joint Computer Conference,* 1953, pp. 53–57; David A. Anderton, Project Typhoon Aids Missile Designers, *Aviation Week,* Dec. 18, 1950, pp. 40–41. Project Hurricane also intended to simulate ballistic missile flights, but that was not its main purpose.

2   Massachusetts Institute of Technology, Servomechanisms Laboratory, Project Whirlwind, Summary report no. 1, Apr. 1946; National Air and Space Museum, National Bureau of Standards Archive 1-8. See also Kent C. Redmond and Thomas M. Smith, *Project Whirlwind: The History of a Pioneer Computer* (Bedford, Mass.: Digital Press, 1980).

3   MIT, Servomechanisms Lab, Project Whirlwind, Summary report no. 2, Nov. 1947, p. 4.

4   Redmond and Smith, *Project Whirlwind,* pp. 156–157.

5   Martin Weik, A Survery of Domestic Electronic Computing Systems, U.S. Army, Ballistic Research Laboratory, Aberdeen, Md., report no. 971, Dec. 1955, pp. 195–196. See also Emerson Pugh, *Memories That Shaped an Industry: Decisions Leading to IBM System/360* (Cambridge: MIT Press, 1984), p. 81.

6   Weik (1955) lists only a few commercial computers available with core memory; the next edition of his *Survey,* published by the Ballistic Research Laboratory two years later, shows that core memory dominated commercial computers.

7   Emerson Pugh, *Memories That Shaped an Industry,* pp. 240–248. See also Claude Baum, *The System Builders: The Story of SDC* (Santa Monica: System Development Corporation, 1981), pp. 12–13.

8   Charles J. Bashe, Lyle R. Johnson, John H. Palmer, and Emerson Pugh, *IBM's Early Computers* (Cambridge, Mass.: MIT Press, 1986), pp. 513–519.

Chapter 6

1   Especially relevant to this story is a comparison with the debate over large versus small dams for flood control and the generation of hydroelectric power. That debate raged in the halls of Congress and at the offices of the U.S. Bureau of Reclamation and Army Corps of Engineers during the 1940s and 1950s. The big-dam interests won out. The result is a series of very large dams on every major river of the Missouri watershed.

2   For an example of how bigness won out in electric power enginering, see Thomas Parke Hughes, *Networks of Power: Electrification in Western Society 1880–1930* (Baltimore: Johns Hopkins, 1983). My discussion of the invention of the integrated circuit is based primarily on Ernest Braun and Stuart Macdonald, *Revolution in Miniature: The History and Impact of Semiconductor Electronics* 2nd ed., (Cambridge: Cambridge University Press, 1982).

3   Simon Ramo, The Impact of Missiles and Space on Electronics, *Proc. IRE,* May 1962, pp. 1237–1241.

4   A case of particular relevance is that of weapons designers, who in the mid-1950s succeeded in making hydrogen bombs compact and lightweight, a breakthrough that helped tip the balance in favor of ballistic missiles as a strategic weapon. Ironically, as Ramo points out, the Soviets lagged in this area, so they made a greater effort in

designing rocket boosters with a heavier lift capacity. That gave them a lead in rocket-booster capability, which the United States did not surpass until the Saturn program in the mid-1960s.

5   Herman O. Stekler, *The Structure and Performance of the Aerospace Industry*, (Berkeley: University of California Press, 1965), chapter 2. See also USAF Investigates Basic Molectronics, *Aviation Week*, August 17, 1959, pp. 77–81.

6   Einstein postulated this as part of his general theory of relativity. There is indeed no way to distinguish between the two types of acceleration, and this was the basis for some physicists' skepticism.

7   James S. Farrior, Inertial Guidance: Its Evolution and Future Potential, in Ernest Stuhlinger, Frederick I. Ordway III, Jerry McCall, and George C. Bucher, eds., *Astronautical Engineering and Science: Honoring the Fiftieth Birthday of Wernher von Braun* (New York: McGraw-Hill, 1963), pp. 148–158.

8   Richard C. Nelson, ed., *Government and Technical Progress, A Cross-Industry Analysis* (New York: Pergammon, 1982), pp. 90–100. The initial Block I Apollo computer had discrete transistors, but for the more advanced Block II computer, used in actual Lunar missions, integrated circuits were used.

9   Braun and Macdonald, *Revolution in Miniature*, chapter 8.

10   Braun and Macdonald (*Revolution in Miniature*, chapter 8) describe some of the other techniques in competition. One was Tinkertoy, a project of the National Bureau of Standards and U.S. Navy. Amazingly it used vacuum tubes instead of transistors (see also Whatever Happened to Project Tinkertoy? *IEEE Spectrum*, May 1987, p. 20). A similar approach was called Micromodule, an Army Signal Corps project. Meanwhile, the Air Force had embarked on something they called Molecular Electronics, a scheme by which they hoped to get individual molecules of material to perform switching (see USAF Investigates Basic Molectronics, *Aviation Week*, August 17, 1959, pp. 77–81). The IBM 360 computer, announced in 1964 and one of the most influential commercial machines ever, used a method whereby small numbers of circuits were laid down on a ceramic base, a technique IBM called Solid Logic Technology.

11   Braun and MacDonald, *Revolution in Miniature*, p. 98.

12   Jack Kilby, Invention of the Integrated Circuit, *IEEE Transactions on Electron Devices*, ED-23, no. 7 (July 1976), pp. 648–654.

13   Philip Klass, Minuteman Guidance and Control, *Aviation Week and Space Technology*, Oct. 29, 1962, pp. 57–66 and Nov. 5, 1962, pp. 74–81.

14   An interview with Richard Tanaka, Smithsonian Computer History Project, January 4, 1973.

15   Rockwell International, The Evolution of Minuteman Guidance and Control, report X75-528/201, June 11, 1975, Autonetics Group, Anaheim, California.

16   Fred Kaplan, *The Wizards of Armageddon* (New York: Simon and Schuster, 1983), pp. 278–283. Kaplan calls the concept a controlled response in this book, although it was known in the Air Force by the other term.

17   Besides upgrading the computer's processor, Autonetics also developed a method of storing programs on a magnetic disk similar to those found on personal computers a decade later.

18   Rockwell International, Evolution of Minuteman Guidance and Control, report no. X75-528/201, June 11, 1975, Autonetics Group, Anaheim, California, p. 8.

19   Minuteman Is Top Semiconductor User, *Aviation Week*, July 26, 1965. See also P. E. Haggerty, The Economic Impact of Integrated Circuitry, *IEEE Spectrum*, June 1964, pp. 80–82 and P. E. Haggerty, Integrated Electronics—A Perspective, *Proceedings IEEE*, 52/12 (1964), pp. 1400–1404. One question relevant to this sequence of events is why the IBM corporation, with its huge internal market for semiconductors, was not a major consumer of integrated circuits at this time. The answer is that IBM, at the time intensely involved with developing its 360 series of computers, had decided to use a hybrid type of miniaturization called Solid Logic Technology and not go with the as yet unproven integrated circuit invented at Texas Instruments and Fairchild. Thus, until the late 1960s IBM was not a major customer or a producer of integrated circuits.

20   See, for example, the March 18, 1977 special issue of *Science* on "The Electronics Revolution." In Intellectual and Economic Fuel for the Electronics Revolution, John Linvill and Lester Hogan argue that Minuteman and IBM, "because of their use volume, were absolutely pivotal in the early establishment of American semiconductor firms. . . . The Minuteman . . . missile system poured hundreds of millions of dollars into the semiconductor industry at a very important time in its history. This money . . . provided funds necessary for the refinements to achieve a high level of reliability for semiconductor devices. All subsequent semiconductor systems benefited from the technological advances with the new levels of electronic reliability" (pp. 1109–1110).

21   The Apollo program was established with one goal: to land a man on the moon and return him safely. But two other programs also used Apollo hardware: Skylab and the Apollo-Soyuz mission. The first used a converted Saturn upper stage for an earth-orbiting space station, the other linked an Apollo Command and Service Module with a

Soviet capsule. Since these missions used Apollo command modules to take men to and from earth orbit, they also used Apollo guidance computers. The Apollo-Soyuz mission in 1975 ended the use of Apollo and Saturn hardware.

22   Charles Stark Draper, The Instrument Laboratory of the Massachusetts Institute of Technology, MIT, Historical Collection, C-5249, Aug. 3, 1979, appendix. See also Charles S. Draper, Engineer, Guided Astronauts to Moon, New York *Times*, July 27, 1987.

23   Roger E. Bilstein, *Stages to Saturn: A Technological History of the Apollo/Saturn Launch Vehicles,* NASA History Series, NASA, SP-4206 (Washington, D.C. 1980), pp. 241–248. Bilstein notes that the inertial navigation system of which this computer was the "brain" was a direct descendant of the first rudimentary inertial systems developed for the V-2 at Peenemünde.

24   James Tomayko, *Computers in Spaceflight: The NASA Experience,* vol. 18, suppl. 3, of Allen Kent and James G. Williams, eds., *Encyclopedia of Computer Science and Technology* (New York: Marcel Dekker, 1987), chapter 2.

25   Eldon Hall, The Apollo Guidance Computer: A Designer's View, Computer Museum *Report,* fall 1982, pp. 2–5.

26   James Tomayko, NASA's Manned Spacecraft Computers, *Annals of the History of Computing,* 7/1 (1985), pp. 7–18.

27   Tomayko, NASA's Manned Spacecraft Computers, and Eldon Hall, The Apollo Guidance Computer: A Designer's View, Computer Museum *Report,* fall 1982, pp. 2–5.

28   A. Michal McMahon, The Computer and the Complex: A Study of Technical Innovation in Postwar America (unpublished manuscript, NASA History Office, Washington, D.C., 1986).

29   Eldon Hall, The Apollo Guidance Computer, and an interview with Hall, April 7, 1986.

30   Tomayko claims that Apollo prototype development consumed sixty percent of the total U.S. output of integrated circuits (NASA's Manned Spacecraft Computers, pp. 33–34).

31   Evan Herbert, Apollo: The Driver and the Driven, *IEEE Spectrum,* Sept. 1983, pp. 56–58. Herbert's view is echoed by C. Gordon Bell, for years the chief computer designer at Digital Equipment Corporation. In an interview conducted in 1985 he regarded chip technology as "the driving devil," "conspiring" to push a product into the marketplace. *Computerworld,* Oct. 14., 1985, p. 19.

32   Deborah Shapley, Technology Creep and the Arms Race: ICBM Problem a Sleeper, *Science,* 201 (Sept. 22, 1978), pp. 1102–1105.

33   The RCA Spectra Series of mainframes, announced in 1965 as direct replacements for IBM's 360 series of computers, was the first commercial mainframe computer to use integrated circuits. That same year Scientific Data Systems delivered their SDS-92 computer, which also used integrated circuits. DEC's PDP-8 has been called by its inventors the first true minicomputer because it was the first computer small enough to fit on a standard equipment rack. Data General's Nova, announced in 1969, was the first to use medium-scale integrated circuits.

Chapter 7

1   Edwin T. Layton, Jr., Technology as Knowledge, *Technology and Culture,* 15 (Jan. 1974), pp. 31–41. The aerospace historian Walter Vincenti has suggested that we add to Layton's focus on design an equal focus on production (examined in this chapter as aerospace manufacturing). See Walter G. Vincenti, Technological Knowledge without Science: The Innovation of Flush Riveting in American Airplanes, ca. 1930–ca. 1950, *Technology and Culture,* 25 (July 1984), pp. 540–576.

2   In some shops this is called an engineering change notice.

3   I wish to thank William Howard of Washington, D.C., for supplying material used in this section.

4   This account is deliberately oversimplified and dwells on obvious and visible differences that a commercial aviation passenger can notice. See Lawrence K. Loftin, *Quest for Performance: The Evolution of Modern Aircraft* (Washington, D.C.: NASA, SP-468), 1985.

5   For an intriguing look at finite element analysis, see Karen A. Frenkel, Computers, Complexity, and the Statue of Liberty Restoration, *Communications of the ACM,* 29/4 (1986), pp. 284–296.

6   From the late 1920s data for some wing cross sections were stored, as were data for the famous set of NACA airfoils, established by the National Advisory Committee for Aeronautics, now NASA. For a discussion of the NACA Standard Airfoils, see James Hansen, *Engineer in Charge: A History of the Langley Aeronautical Laboratory, 1917–1958* (Washington, D.C.: NASA, 1987), pp. 97–98.

7   An equation in which no term is squared, cubed, etc., produces a straight line. A quadratic equation, in which terms are squared, produces a circle or an ellipse. Cubic, quartic, quintic, and higher-order equations produce lines of more complicated curvature: the higher the order, the more complex the curve. With traditional methods a designer could work with cubic equations only with great effort; any equations of

higher order are out of the question. Computers can draw equations of just about any order; however, the complexities of fitting these curves into a design have meant that quintic equations are about as high as anyone wants to go right now.

8   See, for example, Walter Vincenti, Technological Knowledge without Science. See also David Noble, *Forces of Production: A Social History of Industrial Automation* (New York: Alfred A. Knopf, 1984).

9   The Robotic Industries Association, an American trade association of manufacturers and users of industrial robots, has defined "robot" as "a reprogrammable, multifunctional manipulator designed to move material, parts, tools, or specialized devices through variable programmed motions for the performance of a variety of tasks." The word is due to the Czech playright Karl Capek, who used it in his 1921 play *R. U. R.* The word is derived from the Czech *robota,* meaning forced laborers. In the play the lead character, Rossum, manufactures robot workers, who eventually revolt and kill their master.

10   As a typical example, McDonnell-Douglas at its Long Beach, California, plant turns out two and a half medium-range transport planes a week.

11   For a discussion of the difficulties of adapting automobile production techniques to aircraft production, see I. B. Holley, A Detroit Dream of Mass-Produced Aircraft: The XP-75 Fiasco, *Technology and Culture,* 28 (July 1987), pp. 578–593.

12   For a detailed account of the history of numerical control see David F. Noble, *Forces of Production.* Noble forcefully argues that numerical control did not and by nature cannot replace the skills of the human machinist, but that it was forced upon the industry by strong Air Force pressure.

13   This high pitch of activity is captured well in a special issue of *Scientific American* (September 1952) on automatic control.

14   Douglas T. Ross, APT Session, in Richard Wexelblat, ed., *History of Programming Languages* (New York: Academic Press, 1961), pp. 279–367.

15   David F. Noble, *Forces of Production,* p. 208.

16   Herb Brody, CAD Meets CAM, *High Technology,* May 1987, pp. 12–18.

17   The physics community shares an apocryphal story of how in the 1920s Arnold Sommerfeld, a leading German theoretical physicist, remarked that when he died and met his maker, he would ask him to explain two phenomena: turbulent fluid flow and the inner structure of the atom. By the time of his death in 1951 advances in quantum

mechanics had opened up the atom to our understanding, so then it would have been unnecessary to ask about the atom. But in 1951 the physics of fluid flow was still full of unknowns, and Sommerfeld's question was a legitimate one.

18   George Stokes, On the Theories of the Internal Friction of Fluids in Motion, and of the Equilibrium and Motion of Elastic Solids, *Cambridge Philosophical Transactions*, 8, pt. 3, p. 2.

19   Anyone who has watched low-budget Hollywood science-fiction or adventure movies has seen the problems of making a model accurate: such movies typically portray tidal waves, earthquakes, floods, and naval battles using scale water tanks and ships. The behavior of the water usually gives the technique away, because the water's viscosity does not scale up in the same proportion as its wave action.

20   This is usually expressed in terms of a quantity called a Reynolds number, a dimensionless number that represents relationships among the fluid's pressure, density, and so forth. If one can create wind-tunnel conditions with the same Reynolds number as conditions of the actual aircraft, the results should be useful despite the different physical scale of the model.

21   Donald D. Baals and William R. Corliss, *The Wind Tunnels of NASA* (NASA, publication no. SP 440, 1981). See also Alex Roland, *Model Research: The National Advisory Committee for Aeronautics 1915–1958* (NASA, SP-4103, 1985), especially chapter 4, ''Tunnel Vision.''

22   James C. Malin, *The Contriving Brain as the Pivot of History* (Lawrence, Kansas, 1959).

Chapter 8

1   Tony LeVier, *Pilot* (New York: Harper and Brothers, 1954), p. 9.

2   Michael Collins, *Carrying the Fire* (New York: Ballantine Books, 1974), p. 13.

3   The supercritical wing is in fact relatively old, having first been described in theory by Richard T. Whitcomb in the 1960s. Its advantage is that it does not offer the steep increase in drag that most wings do when approaching the speed of sound. Since the first airplane that flew with such a wing in 1971, a variant of the supercritical wing has been used on a variety of large aircraft. The X-29 is the first airplane to employ this shape in a small, light wing with a thin cross section. For more on this concept, see Baals, *The Wind Tunnels of NASA*.

4   This comment is heard at the postflight press conferences of nearly every shuttle mission; John Glenn also made this comment after his earth-orbital flight in 1962. One time the astronauts did *not* say it was after Apollo 11, the first manned landing on the

moon. That flight had more than its share of surprises and unknowns, especially during the last few hundred feet of descent onto the lunar surface. Without the extensive training the astronauts had on simulators, the mission would not have succeeded, but a lot of what happened was unanticipated.

5   Howard S. Wolko, ed., *The Wright Flyer: An Engineering Perspective* (Washington, D.C.: Smithsonian Institution Press, 1987), especially chapter 2, Aerodynamics, Stability, and Control of the 1903 Wright Flyer, by F. E. C. Culick and Henry R. Jex. The authors note that "The most fundamental aspect of the Wrights' invention of the airplane was the idea of the need for control of both roll and yaw motions. It is the foundation of their basic patent submitted in 1902 and granted in 1906" (p. 34). By connecting the wing warp with the rudder cables, the 1903 flyer allowed one action by the pilot to control both roll and yaw during a turn. In this fashion the Wrights were able to manage with one control three of the six degrees of freedom: yaw, roll, and heading. A forward-mounted stabilizer controlled pitch (and indirectly altitude), while the engine throttle controlled airspeed (and, with the stabilizer, altitude).

6   Thomas Parke Hughes, *Elmer Sperry: Inventor and Engineer* (Baltimore: Johns Hopkins University Press, 1971).

7   A two-wheeled bicycle is of course inherently unstable and will crash without constant attention from its rider. Walter Vincenti points out that children learn to ride with training wheels, which give the bicycle inherent stability. But as soon as they master the technique, they gladly sacrifice the stability of the training wheels for the freedom allowed by the unstable configuration.

8   Howard Wolko, ed., *The Wright Flyer: An Engineering Perspective*, especially pp. 19–78; and Walter G. Vincenti, How Did It Become "Obvious" That an Airplane Should Be Inherently Stable? *American Heritage of Invention and Technology*, 4/1 (spring–summer 1988), pp. 50–56.

9   A canard mounted close to the wing, as it is on the X-29, offers another advantage. The vortex of turbulent air that trails behind the canard flows immediately over the wing and thus makes the wing more effective over a wider range of air speeds.

10   James Tomayko, Digital Fly-by-Wire: A Case of Bidirectional Technology Transfer, *Aerospace Historian*, March 1966, pp. 10–18.

11   Columbia Lands with Spacelab Despite Computer Malfunctions, *Aviation Week and Space Technology*, Dec. 12, 1983, p. 23.

Chapter 9

1   Nicholas Wirth, Data Structures and Algorithms, *Scientific American*, Sept. 1984, pp. 60–69.

2   The etymology of "software" is itself an interesting puzzle. One story has it that in 1946 Merrill M. Flood, a mathematician at RAND, coined the term in a report to the War Department to distinguish costs not directly attributable to military hardware. The term apparently stuck despite an admonishment from General Eisenhower, who according to Flood said, "There'll be no software in this man's Army!" See Flood's letter to the editor, *Datamation,* Dec. 1, 1984, pp. 15–16. Dictionary definitions of the term indicate that it came into common use among computing professionals around 1959.

3   This assumption is generally not true; however, in the historical examples cited later in this chapter, the software, especially the Apollo software, was severely constrained by hardware. At present problems of program-execution speed are usually solved by waiting a while until advances in microelectronics solves them at no increase in cost.

4   For the preceding discussion of these themes I have drawn primarily from the writings of Edsger W. Dijkstra, especially his *Selected Writings on Computing: A Personal Perspective* (New York: Springer-Verlag, 1982).

5   And hence the rise of systems-integration companies like TRW and MITRE, mentioned in chapter 6. These companies write the systems software that ties together all the other contractors' equipment.

6   In this chapter I do not examine the various programming languages that aerospace computer systems use. In many cases those languages are no different from those used elsewhere: Fortran, BASIC, C, and others. In some cases the aerospace community has developed its own languages especially well suited to its needs: APT, a manufacturing language; JOVIAL, developed at System Development Corporation and now used for the U.S. Air Traffic Control system; and Ada, mandated by the Department of Defense for all future military systems.

7   In 1946 Grace Murray Hopper, a Navy officer assigned to the Navy's Dahlgren Proving Ground, found that a moth trapped in one of the relays caused the Navy's Mark II computer to fail. She taped the moth in the computer's logbook and appended a note saying that she had found the "bug" that was causing the problem. This incident is often described as the first computer bug; if so, it ironically was also one of the last times that "bug" referred to a hardware and not a software problem. The full story is told by Hopper in *Annals of the History of Computing,* 3 (1981), p. 285.

8   Edward J. Joyce, The Art of Space Software, *Datamation,* Nov. 1985, pp. 30–34. One systems programmer describes many of these notes as falling into the category of "Please don't eat the daisies," which reflects not so much a bug as a basic philosophical difference between user and programmer.

9   Note that the media that carry software (disks, tapes, punched cards, etc.) are physical, but they are not the software. Software engineering is concerned only with the codes that those media carry.

10   Clayton Koppes, *JPL and the American Space Program: A History of the Jet Propulsion Laboratory* (New Haven: Yale University Press, 1982), p. 128. See also Mariner Spacecraft (News Release No. 62-182, Aug. 15, 1962, NASA Headquarters, Washington, D.C.).

11   An interview with John Norton, October 1986 and January 1987, and a videotaped interview with General Jack Albert, the Air Force liaison with NASA for the Agena upper stage on the Mariner program, December 9, 1987.

12   The Mariner I failure was thus a *combination* of a hardware failure and the software bug. The same flawed program had been used in several earlier Ranger launches with no ill effects.

13   A common variant of the story is to describe the error as writing the Fortran statement "DO 3 I = 1, 3" incorrectly as "DO 3 I = 1.3". That caused a new and bogus variable "DO3I" to be assigned the erroneous value 1.3 (many Fortran compilers do not recognize spaces). Since the Atlas Guidance Computer did not have a Fortran compiler, this mistake, if ever made, had nothing to do with Mariner I.

14   These two lines of program code, while they tell us a lot about the nature of software, have no connection with the Mariner I incident. The error was in the original equations, not in the coding, and in any case, the Atlas Launch computer did not even use the Fortran programming language. How the story has become embellished in this way is a mystery. *Annals of the History of Computing*, 6 (Jan. 1984), pp. 61–62. Antony Hoare mentions this story in his 1980 ACM Turing Award Lecture, published in *Communications of the ACM*, 24/2 (1981), p. 77.

15   These bugs are frequently cited in debates over the feasibility of the current U.S. Strategic Defense Initiative (SDI), a program of defense against enemy ballistic missiles that will require increases of several orders of magnitude in computer hardware and software capabilities. See David Parnas, Software Aspects of Strategic Defense Systems, *Communications of the ACM*, 28/12 (Dec. 1985), pp. 1326–1335, and subsequent letters to the editor April 1986, pp. 262–265; July 1986, pp. 591–592; Sept. 1986, pp. 830–831, and Oct. 1986, pp. 930–931.

16   For a more complete and documented account of the reliability of missile-defense systems, see Alan Borning, Computer Reliability and Nuclear War, *Communications of the ACM*, 30 (Feb. 1987), pp. 112–131.

17   This theory has also been supported by Seymour Hersh in his book *The Target Is Destroyed* (New York: Random House, 1986).

18   A letter to the editor by Borning from a computer network message sent on Oct. 18, 1984. The network gives the sender's electronic mail address (''Borning@Washington'') but no other information. It was probably sent by Alan Borning, a computer scientist with many years of experience in aerospace systems (see footnote 16 above). The generation and spread of information via these networks is a topic worthy of study all on its own. It is important to note that information spread on such networks is generally *not* subjected to the rigorous editorial controls that govern what appears in print in a scholarly journal or even a daily newspaper. The erroneous story about Mariner I, for example, has been spread largely in this way, and it seems impossible to ever correct that version.

19   Scott Kim, quoted in Susan Lammers, ed., *Programmers at Work* (New York: Harper and Row, 1986), p. 278.

20   For example, doctors believed that swallowing would be impossible in the weightlessness of space. Therefore, astronauts might die of thirst or malnutrition on a long space journey.

21   The phrase was made famous by Tom Wolfe's book, and later the movie, *The Right Stuff.* Wolfe's book must be read with a few grains of salt. Among other things, the phrase ''the right stuff'' was never common among test pilots or astronauts. But Wolfe is correct in that the astronauts had to lobby for any measure of control over the Mercury spacecraft, against the wishes of some (but not all) engineers. The astronauts had less contempt for the engineers, with whom they had all worked as pilots, than with the medical people assigned to the program. This latter group was far more afraid to trust men to do any jobs in space. They had predicted that all sorts of dire (and in hindsight ludicrous) things would happen to a man as soon as he escaped earth's gravity.

22   For a thorough discussion of these issues and of the current design philosophy among people designing systems for air and space craft, see Alan B. Chambers and David C. Nagel, Pilots of the Future: Human or Computer? *Communications of the ACM,* 28/11 (1985), pp. 1187–1199.

23   Lloyd S. Swenson, James M. Grimwood, and Charles C. Alexander, *This New Ocean: A History of Project Mercury* (Washington: NASA, 1966), pp. 500–501. The quote from President Kennedy is found in *Speeches of the President, 1963* (Washington, D.C.).

24   James Tomayko, *Computers in Spaceflight: The NASA Experience,* in *Encyclopedia of Computer Science and Technology,* vol. 18, suppl. 3 (New York: Marcel Dekker, 1987), p. 45. Also see Eldon C. Hall, MIT's Role in Project APOLLO (Final Report on Contracts NAS 9-153 and NAS 9-4065, vol. 3, sec. 4.0; Cambridge, Mass.: Charles Stark Draper Laboratory, August 1972).

25   Computer Overload Laid to Radar Mode, *Aviation Week and Space Technology*, Aug. 4, 1969, pp. 87–89.

26   The rendezvous radar could be set to one of three modes: autotrack, in which it seeks out and locks on to the orbiting Command Module; slew, in which the crew manually turns the antenna to find the Command Module; and Lunar Module Guidance Computer, in which the computer performs the seeking function. Early mission plans had called for it to be set in slew position during descent, but the change to the mission left it in the autotrack mode. See the previous reference.

27   Telephone interview by the author with Steven Bales. See also Courtney Borks, James M. Grimwood, and Lloyd S. Swenson, Jr., *Chariots for Apollo: A History of Manned Lunar Spacecraft* (Washington, D.C.: NASA History Office), 1979, p. 343. It has also been noted that Armstrong's pulse went up to its highest rate throughout the mission just prior to landing, and it seems reasonable to assume that the computer alarms were responsible. In the design and simulation of the landing, it was planned that the astronauts would land with 5 to 6 percent of their fuel left in reserve (this includes twenty seconds of fuel reserved at all times for a potential abort). Armstrong landed with 2.15 percent of the fuel left, or forty-four seconds, of which he could have used only another twenty-four seconds to maneuver across the boulders. See Computer Overload Laid to Radar Mode, *Aviation Week and Space Technology*, Aug. 4, 1969.

28   Barry Boehm, Software and Its Impact: A Quantitative Assessment, *Datamation*, 19 (May 1973), pp. 48–59.

29   The exchange of letters appeared in *Datamation*, Sept. 1973, p. 161.

30   In this text I have not discussed the role of computation as it pertains to the Strategic Defense Initiative (SDI, popularly known as ''Star Wars''). This program will of necessity call on sophisticated computer hardware and software to identify an enemy threat and guide an appropriate response. The basic concepts involved are no different from previously discussed historical examples, especially SAGE. While many of the systems planned for the Strategic Defense Initiative are classified and represent engineering or theoretical breakthroughs in rocketry, lasers, and nuclear weapons, it is safe to assume that planners of the initiative have no computational tricks up their sleeve. Thus, the arguments about systems software and its reliability hold true for Star Wars as well.

31   See, for example, the interview by Steve Olson with Edsgar W. Dijkstra, The Sage of Software, *Science '84*, Jan.–Feb. 1984, pp. 75–80.

Chapter 10

1   Claude Baum, *The System Builders: The Story of SDC* (Santa Monica: System Development Corporation, 1981). See also Barry Boehm, Software and Its Impact: A Quantitative Assessment, *Datamation*, 19/5 (1973), pp. 48–59.

2   It increasingly appears that aerospace needs the computer industry to survive. In recent years, as it has become harder and harder for U.S. Aerospace companies to make a profit, they have found that marketing their computer and electronics expertise is a very profitable enterprise. Boeing Computer Services, McDonnell Douglas's Information Group, and Martin Marietta Data Systems are solidly profitable divisions of their parent companies. If they stood alone, they would rank among the top fifty U.S. computer companies. The computer business too is of course highly competitive. These companies thrive by marketing what they do best, namely, CAD/CAM, systems software, and networking.

3   ''Ada'' is a registered trademark of the Department of Defense. It was named after Ada Augusta, Countess of Lovelace, an English mathematician who helped publicize Charles Babbage's efforts to build his analytical engine. She has been called ''the first programmer,'' but that is too generous: although she did understand how the analytical engine might be programmed, Babbage never finished that machine, so she never had a chance to implement those ideas.

4   Use of Ada (r) in Weapons Systems, Department of Defense, Directive Number 3405.2, March 30, 1987. See also Jean Ichbiah, Ada: Past, Present, Future: An Interview with Jean Ichbiah, the Principal Designer of Ada. *Communications of the ACM*, 27 (1984), 990–1001. Two critiques are Antony Hoare, The Emperor's Old Clothes, ACM 1980 Turing Award Lecture, *Communications of the ACM*, 24 (1981), and a report by the Task Force on Military Software of the Defense Science Board, reported in *Electronics*, Nov. 12, 1987, p. 121.

5   Deborah Shapley, Technology Creep and the Arms Race: ICBM Problem a Sleeper, *Science*, 201 (Sept. 22, 1978), pp. 1102–1105.

Appendix

1   Presidential Commission on the Space Shuttle *Challenger* Accident, *Report to the President,* Washington, D.C., June 6, 1986. See also Michael Schrage, Shuttle's Warning System Missed Fatal Clue, *Washington Post*, Feb. 3, 1986, p. A9; and Schrage, Rechecking Shards of *Challenger*'s Nervous System, *ibid.*, Jan. 30, 1986, p. A16.

2   David E. Sanger, Shuttle Experts Doubt Computers Could Detect Fire, New York *Times*, Feb. 2, 1986, pp. 1, 18.

3   Richard Feynman, Appendix F—Personal Observations on the Reliability of Shuttle, in Presidential Commission on the Space Shuttle *Challenger* Accident, *Report to the President*, pp. F1–F5.

Bibliography

Aiken, Howard. *A Manual of Operation for the Automatic Sequence Controlled Calculator.* Cambridge: Harvard University Press, 1946. Reprinted by the MIT Press, 1985.

Armer, Paul. SHARE—A Eulogy to Cooperative Effort. *Annals of the History of Computing,* 2 (1980), pp. 122–129.

Aspray, William, and Bruce Brummer. *Guide to the Oral History Collection of the Charles Babbage Institute.* Minneapolis: Charles Babbage Institute, 1986.

Baals, Donald D., and William R. Corliss. *The Wind Tunnels of NASA.* Washington, D.C.: NASA, SP-440.

Bashe, Charles J., Lyle R. Johnson, John H. Palmer, and Emerson Pugh. *IBM's Early Computers.* Cambridge: MIT Press, 1986.

Baum, Claude. *The System Builders: The Story of SDC.* Santa Monica: System Development Corporation, 1981.

Boehm, Barry. Software and Its Impact: A Quantitative Assessment. *Datamation,* 19 (May 1973), pp. 48–59.

Borks, Courtney, James M. Grimwood, and Lloyd S. Swenson, Jr. *Chariots for Apollo: A History of Manned Lunar Spacecraft.* Washington, D.C.: NASA, SP-4205, 1979.

Borning, Alan. Computer System Reliability and Nuclear War. *Communications of the ACM,* 30 (Feb. 1987), pp. 112–131.

Braun, Ernest, and Stuart Macdonald. *Revolution in Miniature: The History and Impact of Semiconductor Electronics.* 2nd ed. Cambridge: Cambridge University Press, 1982.

Brody, Herb. CAD Meets CAM. *High Technology,* May 1987, pp. 12–18.

Bromley, Alan. Charles Babbage's Analytical Engine, 1838. *Annals of the History of Computing,* 4 (July 1982), pp. 196–217.

Ceruzzi, Paul. *Reckoners: The Prehistory of the Digital Computer from Relays to the Stored Program Concept, 1935–1945.* Westport, Conn.: Greenwood Press, 1983.

Chambers, Alan B., and David C. Nagel. Pilots of the Future: Human or Computer? *Communications of the ACM,* 28 (Nov. 1985), pp. 1187–1199.

Charles Babbage Institute for the History of Information Processing. *CBI Newsletter,* 7, no. 1 (fall 1984), p. 6.

Collins, Michael. *Carrying the Fire.* New York: Ballantine Books, 1974.

Dijkstra, Edsger W. *Selected Writings on Computing: A Personal Perspective.* New York: Springer-Verlag, 1982.

Dorfman, Robert. The Discovery of Linear Programming. *Annals of the History of Computing,* 6 (1984), pp. 283–295.

Farrior, James S. Inertial Guidance: Its Evolution and Future Potential. In *Astronautical Engineering and Science: From Peenemuende to Planetary Space, Honoring the Fiftieth Birthday of Wernher von Braun,* edited by Ernest Stuhlinger, Frederick I. Ordway III, Jerry McCall, and George C. Bucher. New York: McGraw-Hill, 1963.

Frenkel, Karen A. Computers, Complexity, and the Statue of Liberty Restoration. *Communications of the ACM,* 29 (1986), pp. 284–296.

Gruenberger, F. J. The History of the Johnniac. *Annals of the History of Computing,* 1 (1979), pp. 49–64.

Hall, Eldon. The Apollo Guidance Computer: A Designer's View. Computer Museum *Report,* fall 1982, pp. 2–5.

Hansen, James. *Engineer in Charge: A History of the Langley Aeronautical Laboratory, 1917–1958.* Washington, D.C.: NASA, 1987.

Herbert, Evan. Apollo: The Driver and the Driven. *IEEE Spectrum,* Sept. 1983, pp. 56–58.

Hersh, Seymour. *The Target Is Destroyed.* New York: Random House, 1986.

Hoag, David G. The History of Apollo Onboard Guidance, Navigation, and Control. *Journal of Guidance, Control, and Dynamics,* 6, no. 1 (Jan.–Feb. 1983), pp. 4–13.

Hoare, Charles Antony Richard. The Emperor's Old Clothes. 1980 ACM Turing Award Lecture. *Communications of the ACM,* 24 (Feb. 1981), pp. 75–83.

Holley, I. B. A Detroit Dream of Mass-Produced Aircraft: The XP-75 Fiasco. *Technology and Culture,* 28 (July 1987), pp. 578–593.

Hopper, Grace Murray. The First Bug. *Annals of the History of Computing*, 3 (July 1981), pp. 285–286.

Howard, William. Interactive Graphics in the F-18 Design. Paper presented at the 1976 Aerospace Engineering and Manufacturing Meeting of the Society of Automotive Engineers, San Diego, Nov. 29–Dec. 2, 1976. Paper MCAIR 76-018, McDonnell Aircraft Company, St. Louis.

Hughes, Thomas Parke. *Elmer Sperry: Inventor and Engineer.* Baltimore: Johns Hopkins University Press, 1971.

Hughes, Thomas Parke. *Networks of Power: Electrification in Western Society 1880–1930.* Baltimore: Johns Hopkins University Press, 1983.

Hurd, C. C. Edited Testimony. *Annals of the History of Computing*, 3 (1981), pp. 163–182.

Ichbiah, Jean. Ada: Past, Present, Future; An Interview with Jean Ichbiah, the Principal Designer of Ada. *Communications of the ACM*, 27 (1984), pp. 990–1001.

Johnson, L. R. Installation of a Large Electronic Computer. *Proceedings of the ACM Meeting,* Toronto, Sept. 8–10, 1952, pp. 77–81.

Kaplan, Fred. *The Wizards of Armageddon.* New York: Simon and Schuster, 1983.

Kidwell, Peggy, and Juanita Y. Morris. *Smithsonian Computer History Project: A Combined Index to Oral Histories Open to Readers.* Washington, D.C.: National Museum of American History, 1986.

Koppes, Clayton. *JPL and the American Space Program: A History of the Jet Propulsion Laboratory.* New Haven: Yale University Press, 1982.

Lammers, Susan, ed. *Programmers at Work.* New York: Harper and Row, 1986.

Layton, Edwin T., Jr. Technology as Knowledge. *Technology and Culture*, 15 (Jan. 1974), pp. 31–34.

LeVier, Tony. *Pilot.* New York: Harper and Brothers, 1954.

Loftin, Lawrence K. *Quest for Performance: The Evolution of Modern Aircraft.* Washington, D.C.: NASA, SP-468, 1985.

Loveman, Bernard. Reliability of a Large REAC Installation. *Proceedings, Eastern Joint Computer Conference*, 1953, pp. 53–57.

McDougall, Walter A. *The Heavens and the Earth: A Political History of the Space Age*. New York: Basic Books, 1985.

McPherson, James L. Census Experience Operating a UNIVAC System. *Symposium on Managerial Aspects of Digital Computer Installations*. Washington, D.C.: U.S. Office of Naval Research, 1953, pp. 30–36.

Noble, David F. *Forces of Production: A Social History of Industrial Automation*. New York: Alfred A. Knopf, 1984.

Ordway, Frederick III, and Mitchell R. Sharpe. *The Rocket Team: From the V-2 to the Saturn Moon Rocket*. Cambridge: MIT Press, 1979.

Parnas, David. Software Aspects of Strategic Defense Systems. *Communications of the ACM*, 28 (Dec. 1985), 1326–1335.

Presidential Commission on the Space Shuttle *Challenger* Accident. *Report to the President*, June 6, 1986. Washington, D.C.: Office of the President.

Pugh, Emerson. *Memories that Shaped an Industry: Decisions Leading to IBM System/ 360*. Cambridge: MIT Press, 1984.

Ramo, Simon. The Impact of Missiles and Space on Electronics. *Institute of Radio Engineers, Proceedings*. May 1962, pp. 1237–1241.

Randell, Brian, ed. *The Origins of Digital Computers: Selected Papers*. 2nd ed. New York: Springer-Verlag, 1975.

Redmond, Kent C., and Thomas M. Smith. *Project Whirlwind: The History of a Pioneer Computer*. Bedford, Mass.: Digital Press, 1980.

Rees, Mina. The Computing Program of the Office of Naval Research, 1946–1953. *Annals of the History of Computing*, 4, no. 2 (1982), 102–120.

Roland, Alex. *Model Research: The National Advisory Committee for Aeronautics, 1915–1958*. Washington, D.C.: NASA, 1985.

Ross, Douglas T. APT Session. In *History of Programming Languages,* edited by Richard Wexelblat. New York: Academic Press, 1981.

Shapley, Deborah. Technology Creep and the Arms Race: ICBM Problem a Sleeper. *Science*, 201 (Sept. 22, 1978), pp. 1102–1105.

Sprague, Richard E. A Western View of Computer History. *Communications of the ACM,* 15 (July 1972), 686–692.

Steckler, Herman O. *The Structure and Performance of the Aerospace Industry*. Berkeley: University of California Press, 1965.

Stern, Nancy. From ENIAC to UNIVAC: An Appraisal of the Eckert-Mauchly Computers. Bedford, Mass.: Digital Press, 1981.

Swenson, Lloyd S., James M. Grimwood, and Charles C. Alexander. *This New Ocean: A History of Project Mercury*. Washington, D.C.: NASA, 1966.

Tomayko, James. NASA's Manned Spacecraft Computers. *Annals of the History of Computing*, 7 (Jan. 1985), pp. 7–18.

Tomayko, James. Digital Fly-by-Wire: A Case of Bidirectional Technology Transfer. *Aerospace Historian*, March 1966, pp. 10–18.

Tomayko, James. Computers in Spaceflight: The NASA Experience. In *Encyclopedia of Computer Science and Technology*, edited by Allen Kent and James Williams. Vol. 18, supplement 3. New York: Marcel Dekker, 1987.

Vincenti, Walter. The Air Propeller Tests of W. F. Durand and E. P. Lesley: A Case Study in Technological Methodology. *Technology and Culture*, 20 (1979), pp. 712–751.

Vincenti, Walter. Technological Knowledge without Science: The Innovation of Flush Riveting in American Airplanes, ca. 1930–ca. 1950. *Technology and Culture*, 25 (July 1984), pp. 540–576.

Vincenti, Walter. How Did It Become "Obvious" That an Airplane Should Be Inherently Stable? *American Heritage of Invention and Technology*, 4 (Spring–Summer 1988), pp. 50–56.

Weik, Martin. *A Survey of Domestic Electronic Digital Computing Systems*. Aberdeen, Md.: U.S. Army, Ballistic Research Laboratory, report no. 971, Dec. 1955, pp. 67–80.

Williams, Michael R. *A History of Computing Technology*. Englewood Cliffs, N.J.: Prentice-Hall, 1985.

Wirth, Nicholas. Data Structures and Algorithms. *Scientific American*, Sept. 1984, pp. 60–69.

Wolko, Howard, ed. *The Wright Flyer: An Engineering Perspective*. Washington, D.C.: Smithsonian Institution Press, 1987.

Wooldridge, E. T. *Winged Wonders: The Story of the Flying Wings*. Washington, D.C.: Smithsonian Institution Press, 1983.

Index